AF712622

65139

Geology of Sedimentary Phosphates

DISCARDED
QE 389.64 .S5 1986

Geology of Sedimentary Phosphates

Maurice Slansky

Bureau de Recherches
Géologiques et Minières

Translated by
Peter Cooper

Preface by
Arthur Notholt

LIBRARY
Appalachian State University
Boone, North Carolina
DISCARDED

Elsevier
New York • Amsterdam • Oxford

English translation © 1986 North Oxford Academic
Publishers Ltd

Original French language edition
(Géologie des phosphates sédimentaires, BRGM) ©1980
Bureau de Recherches Géologiques et Minières

Revised and updated 1986

All rights reserved. No part of this book may be reproduced or transmitted in any form or by any means electronic or mechanical, including photocopying, recording or by any information storage and retrieval system, without permission in writing from the publisher.

English edition first published 1986 by North Oxford Academic Publishers Ltd, a subsidiary of Kogan Page Ltd, 120 Pentonville Road, London N1 9JN in association with Bureau de Recherches Géologiques et Minères, BP 6009, 45060 Orléans Cedex, France.

Sole distributors in the United States and Canada:
ELSEVIER SCIENCE PUBLISHING Co., INC.
52 Vanderbilt Avenue, New York, New York 10017

ISBN 0-444-01073-4

Printed and bound in Great Britain

CONTENTS

Foreword 9

Preface 11

Chapter 1: Distribution of phosphorus in nature 13
1. Historical note 13
2. Phosphorus in the living world 13
3. Phosphorus in the Earth's crust 17

Chapter 2: Mineralogy of sedimentary phosphates 19
1. The sedimentary apatites 19
1.1. Principal substitutions 19
1.2. Other frequent substitutions 24
1.3. Structure of sedimentary apatites 24
1.4. Relations between structure and substitutions 25
1.4.1. Influence of structure on substitutions 26
1.4.2. Influence of substitutions on structure 28
1.4.3. Evaluation of substitutions using crystallographic parameters 31
1.5. Influence of substitution of PO_4 by CO_3 on physical properties 32
2. Other phosphate minerals associated with sedimentary deposits 33

Chapter 3: Petrography of sedimentary phosphates 37
1. Petrographic classification 37
1.1. Definitions 37
1.1.1. Phosphate rocks 37
1.1.2. Phosphate particles 39
1.1.3. Crystalline textures 41
1.2. Phosphatite nomenclature ($P_2O_5 > 18\%$) 42
1.3. Classification of phosphatites 44
1.4. Other phosphorites 46
2. Forms of the phosphate fraction of phosphatites 46

2.1. The microsphatites 46
2.2. Particles 47
2.2.1. Particles of arenite size 50
2.2.2. Particles of rudite size 56
3. The exogangue 61

Chapter 4: Chemistry of phosphatites 69
1. Major elements 69
2. Trace elements 73
3. Concentration of uranium in phosphatites 75
3.1. Most frequent concentrations 75
3.2. Higher concentrations 77
3.3. Types of association between uranium and phosphatites 79
4. Rare earth concentrations in phosphatites 80

Chapter 5: Mechanisms of phosphate sedimentation and phosphatite accumulation 83
1. Possible sources of phosphorus 83
1.1. Direct continental supply 83
1.2. Supply from oceanic waters 86
2. Formation of phosphate minerals from marine solutions 88
2.1. Concentration of phosphorus by marine organisms 89
2.2. Environment for the formation of phosphate minerals 93
2.3. Formation of sedimentary apatites 98
2.3.1. Biogenesis 98
2.3.2. Direct precipitation 99
2.3.3. Epigenesis 103
2.4. Factors triggering the formation of sedimentary apatite 103
2.4.1. Suitable supply 104
2.4.2. Suitability of the depositional environment 105
2.5. Formation of phosphate particles 112
3. Formation of ore deposits 120
3.1. Rate of phosphate sedimentation 120
3.2. Influence of the low rate of phosphatite sedimentation on the facies 124
3.3. Periods favourable to phosphate deposition 125
3.4. Regions favourable to phosphatite accumulation 131
3.4.1. Abundance of phosphate supply 131
3.4.2. Location of phosphate deposits 134
4. Secondary evolution of phosphate deposits 136
5. Mechanisms of uranium concentration in phosphorites 140
5.1. Primary concentration 140
5.2. Secondary concentration 141
6. Examples of environments favourable to the accumulation of phosphatites 142
6.1. Coastal basin of Gabon 142
6.2. Benin – Togo coastal basin 144

6.3. Mauritania – Senegal – Gambia – Guinea Bissau basin 148
6.4. Moroccan coastal basins 153
6.5. Tertiary basin of Tunisia 154
6.6. The Permian of the Rocky Mountains, U.S.A. 157
6.7. Conclusion 159

Chapter 6: Prospecting and mining of deposits 163
1. Prospecting methods 163
1.1. Determining the favourable geological periods 163
1.2. Identification of favourable zones 165
1.3. Detailed study of favourable zones 165
2. Methods of extraction and processing 166

Chapter 7: The phosphate industry 169
1. Economic data 169
2. World resources and reserves 171
3. Uses of phosphates 174
3.1. Manufacture of fertilizers by acid treatment 174
3.2. Fertilizers for direct application (ground phosphates) 176
3.3. Other phosphate fertilizers 176
3.4. Thermal phosphorus and phosphoric acid 177
3.5. Cattle feed 177
3.6. Polyphosphates 178

Chapter 8: Conclusion 179

Bibliography 181

Index 207

Foreword

Phosphorus is an indispensable and irreplaceable element in the living world, whose need of it is steadily climbing as the population grows. But there is a widening gap between what can be supplied from natural reserves of the pedosphere and the demand, growing equally fast, of the agro-industry, which is becoming increasingly dependent on phosphate fertilizer in agricultural production.

In the face of these increasing needs, as well as fluctuations in supply level and heavy price increases in 1974–1975, research on sedimentary phosphates has grown apace. Work has been done on deposits of different ages, from the Precambrian to the Present, using a broad variety of techniques. This research has produced a lot of information, the distribution and interpretation of which has been assisted by international cooperation which has fortunately increased considerably over the past few years, particularly thanks to the organizers of Project 156 of the UNESCO International Geological Correlation Programme.

The first edition of this book, appearing in French in 1980, was intended to provide a state-of-the-art review in the field of sedimentary phosphates, with the aim of giving geologists working on phosphate deposits an overall view of the latest developments and a good list of references to assist them in their work.

The present second edition, in English, has the same aim. It has been revised in response to readers' reactions to the first edition and to take account of the results of recent work, in particular on phosphorites in today's seas and on organic matter associated with phosphate deposits.

Maurice Slansky
Bureau de Recherches
Géologiques et Minières
Orleans

Preface

It is a privilege and a pleasure to introduce the first English edition of Maurice Slansky's 'Géologie des Phosphates Sédimentaires'. In his comprehensive treatment of the subject, the author, who joined the Bureau de Recherches Géologiques et Minières in 1960 and was also Head of its Geological Department from 1964 until 1976, has drawn on invaluable and considerable practical experience as a phosphate geologist, particularly in connection with studies carried out on the major phosphate deposits in West Africa, as well as on many years of applied research. The result is an authoritative work on a raw material that has proved to be a fascinating field of investigation, both academic and industrial.

Phosphorus is indispensable to all forms of animal life and, in a similar context, has for many years been recognized as one of the three major elements essential to plant growth, the other two being potassium and nitrogen, without which agricultural productivity and, hence, food production cannot be maintained or increased. Indeed, the significance of phosphate rock as an inorganic source of the element dates from about the middle of the 19th century, when the German chemist Justus von Liebig demonstrated the role of mineral nutrition in plant metabolism. Since that time, phosphate rock has become the basis of a major industry of worldwide importance, being the only suitable raw material in the manufacture of phosphate fertilizers (by far its most important end use) and the preparation of phosphorus-based chemicals, principally those used in detergents. It is noteworthy also that, in marked contrast to other vital mineral raw materials such as oil, iron ore, copper and sulphur, which can be at least partly replaced by substitutes, there is little, if any, substitution or recycling in the phosphate rock industry.

Phosphate rock is an imprecise term, unfortunately used in the mineral industry for any naturally-occurring raw material that contains one or more phosphate minerals and, after appropriate treatment, possesses chemical and physical characteristics that make it suitable for commercial use. Geologically, therefore, the term encompasses a very wide variety of rock types of diverse origin, character and mode of occurrence. However, deposits of marine, sedimentary origin, the subject of Maurice Slansky's present work, account for most of the world's production and are mainly of Miocene–Pliocene, Upper Cretaceous–Eocene, and Cambrian age. There are large resources and such deposits will undoubtedly continue to provide the bulk of man's phosphate rock requirements, at least for the foreseeable future.

Not unexpectedly, phosphate deposits have been the object of considerable academic and applied research, mainly in response to the stimulus in phosphate exploration, evaluation and commercial development brought about by fluctuations and uncertainties in supply and, notably, by the dramatic increases in phosphate rock prices introduced by major suppliers to world markets during the early and mid-1970s. Particularly important contributions to phosphate geology have been made in recent years, principally those arising from the multidisciplinary effort embodied in the UNESCO-sponsored International Geological Correlation Programme Project 156: Phosphorites. These contributions have highlighted the very wide distribution of sedimentary phosphate rock, both geographically and within the geological column, and the remarkable variety of geological environments in which it can occur. Although differences of opinion remain, there is fairly general agreement that most of the major deposits have formed on stable shelves or platforms in waters usually much less than 500 m deep and in areas of low palaeolatitudes. A trapping mechanism that creates a favourable, restricted, depositional environment is also regarded as necessary for the accumulation of large quantities of phosphate.

The present work on sedimentary phosphates provides readers with considerable insight into all these topics and has drawn attention also to the important role played by micro-organisms in bringing about the formation of anoxic (organic-rich) bottom (sea-floor) sediments deemed so favourable to phosphate precipitation. Much of the available data appeared in the first edition, which was published in French in 1980 and, as such, provided geologists and related research workers with a much needed and comprehensive review of existing geological literature on sedimentary phosphate deposits. The present, updated, English edition has the same basic objectives and is highly recommended as essential reading to all who are interested in the geology and resources of this vital mineral raw material.

Arthur J. G. Notholt
Interfos Associates,
London

Chapter 1

Distribution of phosphorus in nature

1. HISTORICAL NOTE

Phosphorus was extracted from human urine for the first time in 1669 by the German alchemist Henning Brandt, of Hamburg. The same operation was performed in the years that followed by another German alchemist, Johann Kunckel, and by the English chemist Robert Boyle. The name chosen for this substance, which glows in the dark and ignites spontaneously, comes from the Greek word 'phosphoros' (light-bearing). Another century passed before the Swedish chemist Carl Wilhem Scheele discovered the presence of phosphorus in bones, in 1769. A method for extracting phosphorus from the bones of animals was rapidly perfected.

The importance of phosphates for agriculture has been obvious since the end of the 18th century. The first phosphate fertilizers, applied by spreading, consisted solely of crushed bones; guanos arrived on the scene at around the beginning of the 19th century. The existence of natural phosphates and the possibilities that they offered were recognized at the same time, but began to be systematically used only around 1850. The first deposits to be mined were in Europe; these were soon replaced, particularly from the beginning of the 20th century, by the very large deposits discovered in the U.S.A., North Africa and Oceania. Since that time the consumption of phosphates has constantly increased from year to year and, since phosphorus is an indispensable and irreplaceable element of the living world, this growth will inevitably continue.

2. PHOSPHORUS IN THE LIVING WORLD

Phosphorus is an essential constituent of all living cells, where it is found in a mineral or organic form. In mineral form it appears mainly in bone tissue (cf. Table 1, based on McConnell, 1973), generally as calcium phosphate, with 2–2.5% magnesium phosphate. It is, however, also present in all tissue, with a concentration varying from a few milligrams to over a gram of phosphorus per kilogram.

Among the many organic forms of phosphorus present in cells the most important are the nucleic acids, the elementary units of life which combine to

Table 1. *Partial analysis of a scoured cow bone and of enamel from a human tooth* (from McConnel, 1973).

% (weighted)	Bone	Enamel
P_2O_5	28.58	39.90
CaO	37.56	50.93
MgO	0.72	0.35
Na_2O	0.99	0.94
K_2O	0.07	0.037
SiO_2	0.04	0.01
CO_2	3.48	2.08
F	0.07	—
Cl	0.08	0.32
Loss at 68°C	—	2.12
Loss at 500°C	—	2.80

form genes and chromosomes; their fundamental constituents are RNA and DNA (Table 2). These nucleic acids are the depositaries of our genetic inheritance; they guide protein synthesis, are responsible for energy transfer and are involved in many other functions. They contain about 9.5% phosphorus, or approximately 22% P_2O_5 (%P × 2.2914 = %P_2O_5). Adenyl nucleotide phosphates play an equally crucial role, since practically all the bioenergy of living organisms is associated with the presence of ATP, ADP or AMP in cells. ADP contains about 18.5% phosphorus, or over 42% P_2O_5, a proportion comparable with fluorapatite.

Other important organic forms of phosphorus include phosphagene, which plays a major role in muscular contraction, the phospho-amino lipids, cellular constituents of the nervous tissue, bone marrow and reproductive cells, phosphoglyceric acids, which play a role in photosynthesis, the large class of phosphatases, enzymes responsible for the hydrolysis of esters and anhydrides of phosphoric acid, etc.

Phosphorus accounts for nearly 1% of the weight of the human body, corresponding to about 650 g of phosphorus or 1500 g of P_2O_5 on average; 86% of this is in mineral form in bones and teeth. The proportion of phosphorus by weight is of the same order (between 0.8 and 1%) in livestock such as cattle. Their grazing therefore leads to serious impoverishment of the soil in fields where they are bred. For example, a milk cow weighing 500 kg has about 3300 – 3600 g of phosphorus in its bone structure and 680 – 750 g of phosphorus in its soft tissue; it eliminates 0.9 g of phosphorus per kilogram of milk produced, corresponding to 5400 g overall, which is more than its own reserves (Ferrando, 1982).

Table 2. *Various essential organic forms of phosphorus found in nature*

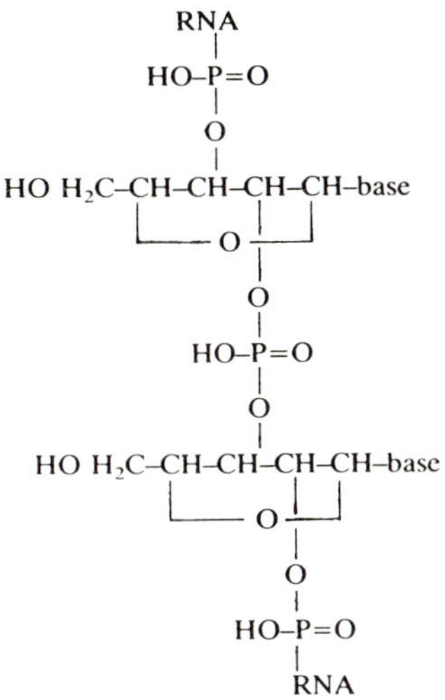

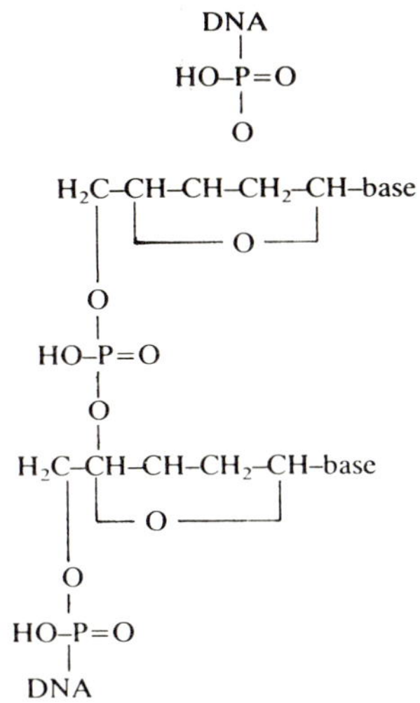

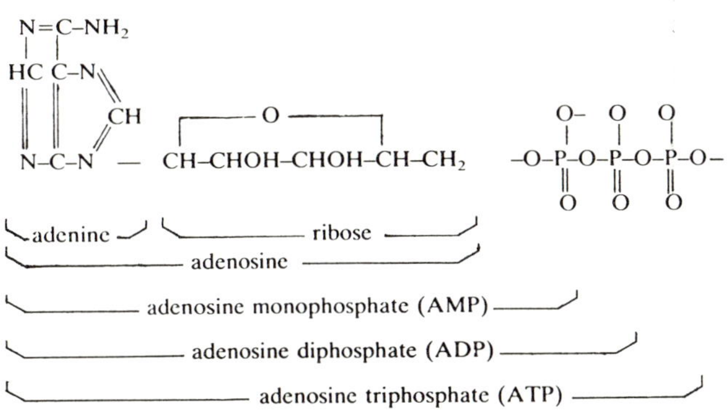

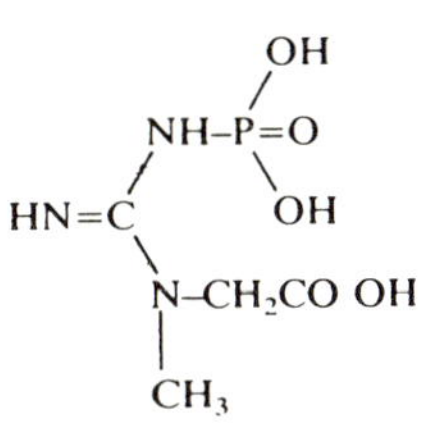

phosphagene

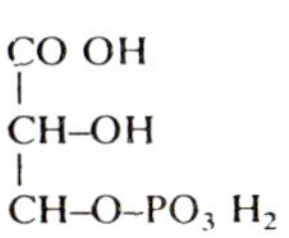

phospho-3-glyceric acid

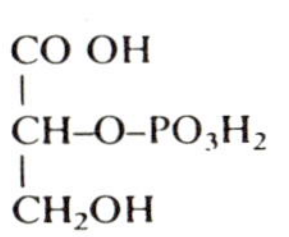

phospho-2-glyceric acid

Having nothing comparable with the requirements associated with bone structure, plants are less rich in phosphorus, though they nevertheless contain significant quantities. The following concentrations, expressed in $\%P_2O_5$ (dry wt.), were found by Gervy (1973), and are accurate to within 15%.

Wheat	Whole plant on efflorescence	1.03
	Grain	1.03
	Straw	0.20
Maize	Grain	0.91
Rice	Whole grain	0.80
	Polished grain	0.45
Asparagus	Stalk	2.06
Haricot beans	Grain	1.37
Tomatoes	Fruit	1.48
Onions	Bulbs	1.25
Mushroooms		2.63
Potatoes	Tubercles	0.45
Soya beans	Grain	1.60

Concentrations of P_2O_5 in marine plants are comparable or, in the case of phytoplankton, even higher. In Cyanophyta (Vinogradov, 1953), for example, dry weight contents of P_2O_5 vary between 0.5 and 1.2%, and can reach as much as 3.7%; concentrations of between 0.7 and 3% have been found in diatoms, and the dinoflagellates (Peridinians) apparently contain similar concentrations.

Among the marine animals, fish contain about 1.8% P_2O_5 (dry wt.) (Bowen, 1966), with concentrations found in their bone structure reaching between 39 and 40%. Crustaceans are generally rich in phosphorus; Clarke and Wheeler (1922) mention *Temora longicornis* and *Thysanoessa inermis*, small plankton species which are very abundant off New England and Newfoundland and whose mineral content, representing 2.77% and 7.68% respectively of their whole weight, is almost exclusively calcium phosphate.

Inarticulate brachiopods, such as Lingula, tend to have a considerable concentration of calcium phosphate in their shells, with a P_2O_5 content in excess of 40%. In the marine Annelids, contents of between 20 and 21% of P_2O_5, as a proportion of their dry weight, have been found for the *Onuphis tubicola* and *Hyalinacia artifex* species. Finally, Bowen finds that the average P_2O_5 content of bacteria is approximately 7%.

3. PHOSPHORUS IN THE EARTH'S CRUST

Ronov and Yaroshevsky (1969) estimate the average content of phosphorus in the Earth's crust to be 0.23% P_2O_5. A large part of this is contained in igneous rocks (the most abundant rock type), in which the average P_2O_5 content varies from 0.16 to 0.42%, according to Nockolds (in McKelvey, 1973). This figure is supported by Table 3, which is based on 695 specimens of French granites and metamorphic rocks.

Where sedimentary rocks are concerned, McKelvey (1973) indicates, from various sources, the following average percentages of P_2O_5:

Sandstones:	0.08 to 0.16
Shales:	0.11 to 0.17
Carbonate rocks:	0.03 to 0.07

At greater depths, carbonate rocks contain 0.08 – 0.18 P_2O_5, argillaceous rocks 0.14 – 0.33, and siliceous rocks 0.13 – 0.27.

The average concentrations of P_2O_5 in the soil and the seas, estimated at 0.09 and 0.02% respectively (Henin, 1982), are often lower still. These low phosphate levels show that plants that draw their nourishment from the soil or

Table 3. *P_2O_5 contents of French metamorphic rocks or granites, based on data held by the B.R.G.M. at Orléans.*

Nature and origin of samples	Number of analyses	Percentage P_2O_5		
		Minimum	Maximum	Mean
Metarenites, metabasites, leptynite-gneiss,	57	0.02	0.32	0.14
epimetamorphic volcanic-detrital rocks	36	0.01	0.37	0.15
from the Bas-Limousin	51	0.01	0.40	0.15
	7	0.05	0.36	0.17
Pyrenees granite	77	0.03	0.60	0.18
	19	0.01	0.26	0.15
	29	0.12	0.44	0.22
Vendée granite	60	0.05	1.53	0.33
Brittany granite	329	0.03	0.30	0.17
	30	0.19	0.37	0.23

water appreciably concentrate the phosphorus they draw from the nutrient solutions.

The formation of phosphate deposits results from an even higher degree of concentration, since their phosphate contents are frequently 100 – 150 times greater than the crustal average. Three types of deposit are mined at present:

– Sedimentary deposits are the most important, both in number and in volume. *In situ* concentrations of P_2O_5 often exceed 20 or even 30%; in the Bou Craa deposit (Morocco), the concentration of P_2O_5 in the ore often exceeds 33 or even 35%. By comparison with the average P_2O_5 concentrations in sandstones, shales and carbonate rocks, those found in sedimentary deposits suggest phosphorus concentration factors often greater than 200 – 300.

– Deposits of igneous origin are associated with intrusive alkaline complexes in which the most abundant rocks are nepheline syenites, carbonatites, ijolites and pyroxenites. These deposits are less common, and often less rich and smaller than sedimentary deposits. Some, however, are very important, such as those at Khibiny (U.S.S.R.) associated with a ring complex of nepheline syenites, containing 1600 million tonnes of ore with an average of 18% P_2O_5. The Palabora deposit (South Africa), of a lower grade, contains equally large reserves of apatite, associated with an intrusion composed of carbonatite and pyroxenite; these reserves are also in part made up of a zone with phoscorite (apatite + serpentine + magnetite) immediately surrounding the carbonatite intrusion.

– Guanos contain smaller reserves, but are nevertheless far from insignificant. The main deposits are due to sea-birds, whose droppings, containing about 4% P_2O_5, react with the underlying bedrock. When this rock is calcareous, the calcium phosphate that forms, which is similar to that of sedimentary deposits, may be sufficiently rich and abundant to justify commercial working: the island of Nauru in the Pacific Ocean, for example, originally had reserves estimated at 90 million tonnes of ore with 39% P_2O_5 when mining began, while Christmas Island in the Indian Ocean had ore reserves and resources containing 23 – 27% P_2O_5 estimated at over 100 million tonnes.

Similarly, the droppings of bats may also be a source of phosphate accumulation, with high grades found in caves. The tonnages formed in this way are, however, far more modest; on the island of Java, for example, where this type of phosphate is frequent, the ore reserves do not appear to exceed a few tens of thousands of tonnes, with a P_2O_5 concentration ranging from 15 to 40%.

Chapter 2

Mineralogy of sedimentary phosphates

The natural phosphates include over 200 mineralogical species (Fischer and Jerome, 1973). The most widespread, however, belong to the apatite family.

In magmatic or metamorphic rocks, the most common apatite is fluorapatite, $Ca_{10}(PO_4)_6F_2$. They may also, though more rarely, contain chlorapatites and hydroxyapatites. These, however, according to McConnel (1973), never appear in their extreme forms – $Ca_{10}(PO_4)_6Cl_2$ and $Ca_{10}(PO_4)_6(OH)_2$: chlorapatites invariably contain fluorine, and hydroxyapatites fluorine or chlorine.

In the sedimentary series it was long thought that calcium phosphates consisted mainly of a fluorine-bearing amorphous material known as collophane or collophanite (or colophanite). Modern mineralogical studies, however, show that this apparently amorphous substance is, in fact, cryptocrystalline or microcrystalline and is made up of minerals of the apatite family.

If it were still in use, therefore, the term 'collophanite', which is petrographic rather than mineralogical, would refer to cryptocrystalline apatitic rocks, but specialists in this field no longer recommend the adoption of the term.

1. THE SEDIMENTARY APATITES

The predominant minerals among sedimentary phosphates are similar to the fluorapatite of igneous rocks but usually differ from them by a set of partial isomorphic substitutions.

1.1. Principal substitutions

The most common substitution in sedimentary apatites is that of PO_4^{3-} by CO_3^{2-}, the introduction of supplementary F^- helping to maintain the electrical neutrality of the crystal. The most frequent mineral is thus carbonate fluorapatite, or francolite, for which we can write a first approximation to a simplified structural formula as follows:

$$Ca_{10}[(PO_4)_{6-x}(CO_3)_x]F_{2+x}$$

The value of x is variable. In 18 specimens studied at the B.R.G.M.[(1)], taken from different deposits (Saudi Arabia, Iran, Brazil, Benin, Senegal, Morocco, India, U.S.A.), it ranges from 0.39 to 1.36, with an average of 0.75. A specimen of Senonian phosphatic chalk from northern France gave a value of 1.42 for x, while an Albian phosphate nodule from the same region gave a value of 1.62; a recent nodule from the Atlantic sea floor was found to have a still higher value of 1.90. On the other hand, a sample from northern Thailand whose concretionary appearance indicates that it has undergone weathering, with some phosphate content returning to solution, takes the form of a fluorapatite without CO_3.

In a note of major importance, Lehr and McClellan (Lehr *et al.*, 1967) give the results of their examination of 107 commercial concentrates obtained from most deposits around the world. The concentrates from deposits of igneous origin have very low values for x[(2)], amounting to between 0.07 and 0.09, whereas in most sedimentary deposits there is a high rate of replacement of PO_4^{3-} by CO_3^{2-}, with x fluctuating widely between 0.4 and 1.4. The Tennessee deposits, produced by the intensive leaching of Ordovician phosphatic limestones, give lower values for x in certain specimens, and the same is true of deposits originating from guano.

Another frequent substitution in sedimentary apatites is that of Ca^{2+} by Na^+ and Mg^{2+}. Lehr *et al.* have shown that this substitution is directly connected with that of PO_4^{3-} by CO_3^{2-}; this relationship is shown in Fig. 1, based on chemical data concerning marine sedimentary deposits published by these authors.

When the molar CO_3/PO_4 ratio reaches 0.3 (x in the above structural formula then being approximately 1.4), about 6% of the Ca^{2+} ions are replaced by Na^+ and Mg^{2+}. It therefore appears that these two elements, in particular Na, which is more abundant and also monovalent, combine with F to preserve the electrical neutrality of the apatite crystal when PO_4^{3-} is replaced by CO_3^{2-}. A correlation between Na content and the degree of substitution of PO_4 by CO_3 had already been noticed by Gulbrandsen (1960, 1966) in U.S. ores.

---

(1) Studied using X-ray diffraction techniques, by C. Jacob.

(2) These authors used a chemical method for determining the CO_2 content of apatites, after elimination of the carbonates.

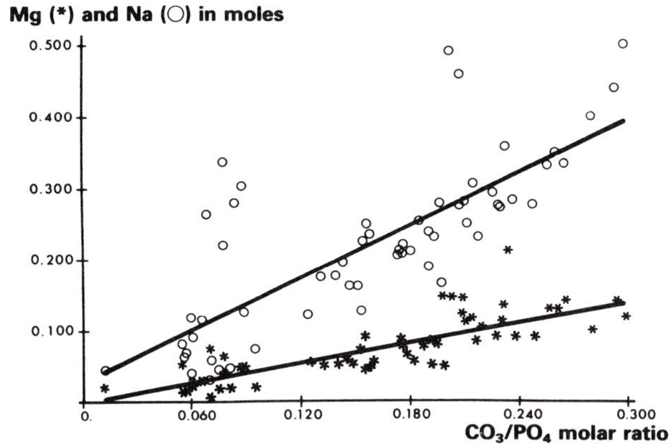

Fig. 1. *Variation in the molar sodium and magnesium contents as a function of PO_4 replacement by CO_3 in apatite* (from Lehr *et al.*, 1967).

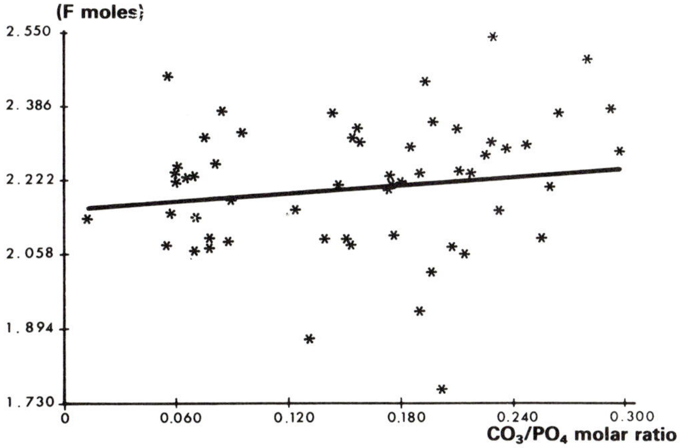

Fig. 2. *Variation in the molar fluorine content as a function of PO_4 replacement by CO_3 in apatite* (from Lehr *et al.*, 1967).

The partial replacement of F by OH is also fairly frequent in sedimentary apatites. The extent to which it takes place can vary. When the proportion of F by weight is below 1% the apatite is called dahllite or carbonate hydroxy-apatite, the structural formula of which may be written $Ca_{10}(PO_4, CO_3OH)_6(OH,F)_2$; it is this substance that provides the mineral content of bones and teeth (with a very low F content). The mineral is very common in phosphates derived from guano, but it is also found, combined with carbonate fluorapatite, among phosphates of marine origin.

This replacement of F by OH is probably responsible for the poor correlation between variations in the molar content of F and the CO_3/PO_4 molar ratio (see Fig. 2).

The possible importance of substitution by Na, Mg or OH, led Lehr *et al.* (1967) to propose the following general structural formula for sedimentary apatites:

$$(Ca,Na,Mg)_{10}(PO_4)_{6-x}(CO_3)_xF_y\ (F,OH)_2$$

in which y varies between 0.33 and 0.5 x, x generally lying between 0 and 1.5. The number of moles of Na, approximately $x-y$, is greater than the number of moles of Mg. The total number of moles of Mg in most marine sedimentary deposits is greater than 2.

1.2. Other frequent substitutions

The sedimentary apatite structure allows a number of other substitutions, though not generally to the same extent as those discussed above. Potassium and strontium, for example, are often present but generally do not occupy more than 2–3$^0/_{00}$ of the calcium sites. Uranium also frequently replaces calcium but in a still lower proportion; the same applies to thorium and the rare earths.

The SO_4^{2-} ion is, according to some authors (Gulbrandsen, 1966), as likely to substitute for PO_4^{3-} as is CO_3^{2-}, while others claim it is far less likely to do so (McClellan and Lehr, 1969). In his study of the Tunisian phosphates, Sassi (1974) shows that the apatites of the Gafsa Basin may contain up to 3.55% by weight of SO_3 with 28% P_2O_5, which corresponds to a SO_4/PO_4 molar ratio of 0.11. This ratio is lower for the deposits situated further north and also in the case of Moroccan specimens (0.06 – 0.10) and a Jordanian specimen (0.05).

Numerous other ions can substitute for PO_4, particularly SiO_4, AsO_4 or VO_4; the extent to which such substitution may take place is not well known. In the laboratory, a partial substitution of PO_4 by AlO_4, associated with the replacement of part of the Ca by Al was carried out by Fischer and McConnell (1969). It sometimes occurs in nature (McConnell, 1973) but the corresponding mineral, abukumalite, has not yet been discovered in sedimentary deposits.

1.3. Structure of sedimentary apatites

The structure of the fluorapatite crystal was established by Naray Szabo (1930) and Mehmel (1931), and various authors have confirmed their basic findings. According to Beevers and McIntyre in particular (in Bonel, 1972 and Young, 1975), fluorapatite, which forms a hexagonal crystal, has a structure that may be described as follows:

– Parallel with reference plane 001, the crystal has two planes of symmetry at levels 1/4 and 3/4. Perpendicular to this reference plane, Ca ions and O atoms alternate along columns parallel to the *c* axis of the crystal. The axes of the columns on which the Ca ions are situated are ternary symmetry axes of the crystal. These axes intersect the 001 plane at points which, when interconnected in pairs by straight lines, form a set of hexagons (Fig. 3a). The Ca ions are mainly situated at levels 0 and 1/2, i.e. half-way between the planes of symmetry; each Ca ion is surrounded by nine O atoms, three in the next highest plane of symmetry, three in the next lowest plane and three in approximately the same plane as the Ca ions (Fig. 3b).

– The columns are interconnected by P atoms situated on the M1/4 and M3/4 planes of symmetry. Each P atom is attached to four O atoms to form a PO_4 tetrahedron; two of the O atoms are in the same plane of symmetry as the P atom, the other two are symmetrically positioned on either side. The two symmetrical O atoms and one of the O atoms in the plane of symmetry belong to one column, and the fourth O atom belongs to the adjacent column (Fig. 3c).

– The Ca – O columns and the PO_4 tetrahedra linking them form a 'behive' network, as can be seen from its projection on the 001 plane (Fig. 3d). The network contains a set of hollow channels perpendicular to that plane. The axis of each channel is a helicoidal senary symmetry axis; the nearest elements of the system are the O atoms which are practically in the same plane as the Ca ions of the columns.

– The channels, however, are not empty. There is an F ion in each plane of symmetry M1/4 and M3/4; this ion is on the axis of each of the channels, at the centre of an equilateral triangle which has further Ca ions at each angle (Figs 3e and 3f). The apatite lattice therefore contains two categories of Ca ions, the Ca_I ions of the columns mentioned before, which are on the ternary axis, and the Ca_{II} ions near the helicoidal senary axes of the channel. The Ca_I and Ca_{II} ions belong to different planes (0 and 1/2 for Ca_I, 1/4 and 3/4 for Ca_{II}).

The unit cell of the fluorapatite lattice is a regular prism with a rhomboid base (Fig. 4). Parameter *a* of the system, the side of the rhombus, is approximately 9.37 Å and parameter *c*, the height of the prism, is approximately 6.88 Å (McConnell, 1973).Each unit cell comprises four Ca_I ions, six Ca_{II} ions, six P atoms, two F ions and 24 O atoms.

1.4. Relations between structure and substitutions

The structure of sedimentary apatites and the substitutions to which they are subject affect each other.

Fig. 3. *Structure of apatite.*

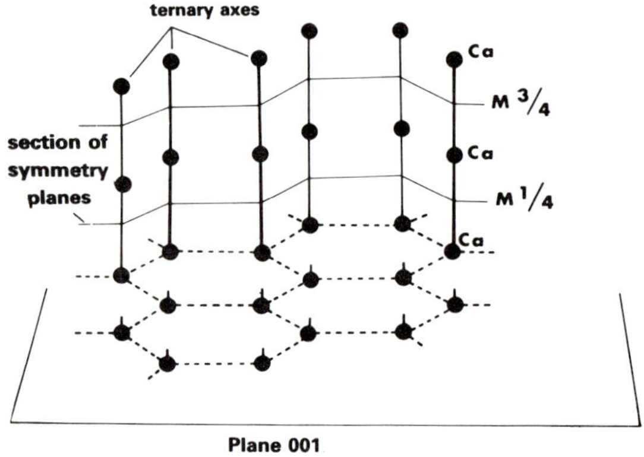

(a) The ternary axes are perpendicular to plane 001 and located at the nodes of linked hexagons. Ca_I ions on these axes are very nearly equidistant between symmetry planes M 1/4 and M 3/4, which are in turn parallel to plane 001.

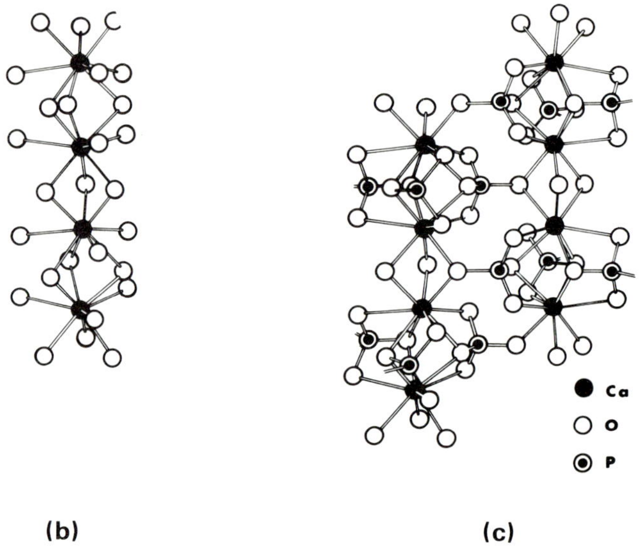

(b) Nine O atoms are associated with each Ca ion on the ternary axes, distributed evenly between the plane of symmetry above the Ca_I ion, the plane of symmetry immediately below, and the plane of the Ca_I ion itself.

(c) The Ca_I – O columns are linked by P ions forming PO_4 tetrahedra with three O atoms from one column and one from the next.

Fig. 3 *Continued*

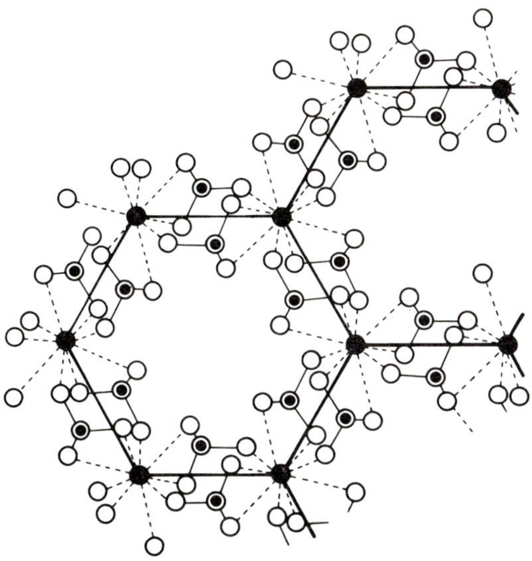

(d) Projection onto plane 001 of the Ca_I – O columns and the PO_4 tetrahedra associated with them.

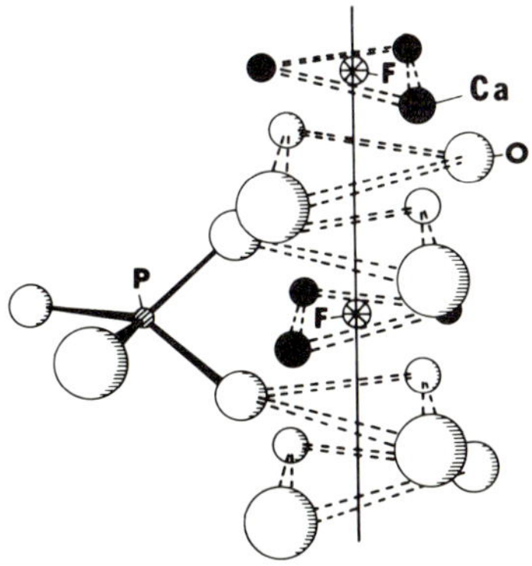

(e) Distribution of atoms around the helicoidal senary axis. F and Ca_{II} are in the M 1/4 and M 3/4 planes of symmetry.

Fig. 3 *Continued*

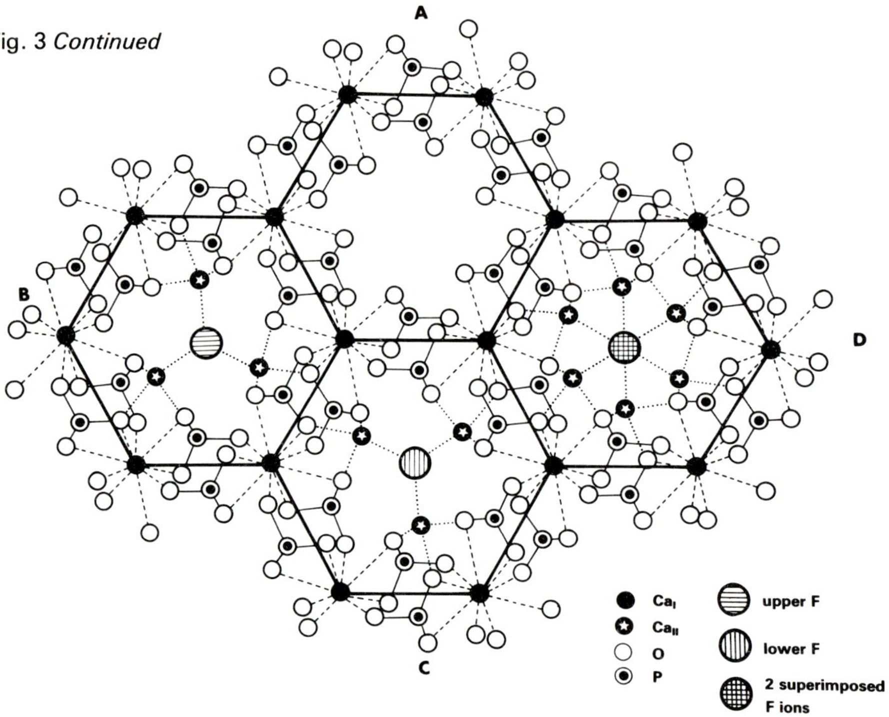

(f) Projection onto plane 001 of the apatite structure (from Altschuler *et al.*, 1958).

— in A: F and Ca_{II} ions have been ignored;
— in B: only F ion is shown;
— in C: the other F ion is shown;
— in D: all ions are shown including the two superimposed F ions and the six Ca_{II} ions.

1.4.1. Influence of structure on substitutions

In the first place, the substitutions obey the usual rules of diadochy: similar ionic radius, maintenance of electrical neutrality, etc. For example, Ca^{2+} (0.99 Å ionic radius) replaced more easily by Mg^{2+}(0.66 Å), Mn^{2+}(0.80 Å) or Na^{+} (0.97 Å) than by larger ions such as Sr^{2+} (1.12 Å), Ba^{2+} (1.32 Å) or K^{+} (1.33 Å); the ionic radius of U^{4+}(0.97 Å) also enables it to replace Ca^{2+}. As far as F^-, OH^- and Cl^- are concerned, their ionic radii (estimated at 1.36, 1.40 and 1.81 Å respectively) are sufficiently similar to allow them to replace each other.

These substitutions are also influenced by the existence of the channels centred on the helicoidal senary axes, within which bonds are looser. These channels, with an average diameter of 3 – 3.5 Å (Montel, 1975), facilitate the

Fig. 4. *Unit-cell of the apatite lattice.*

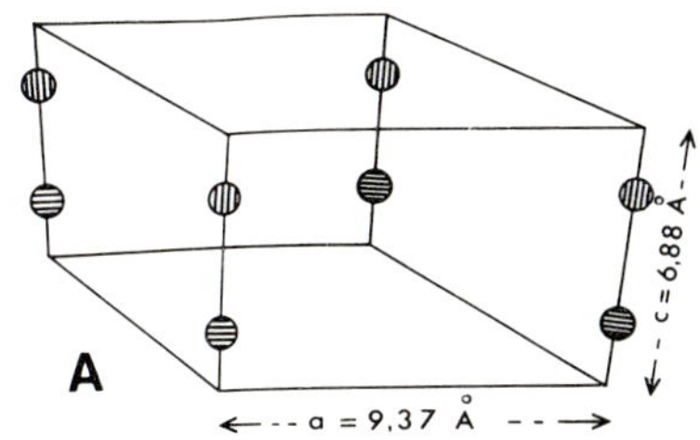

(a) Perspective view showing F ions alone.

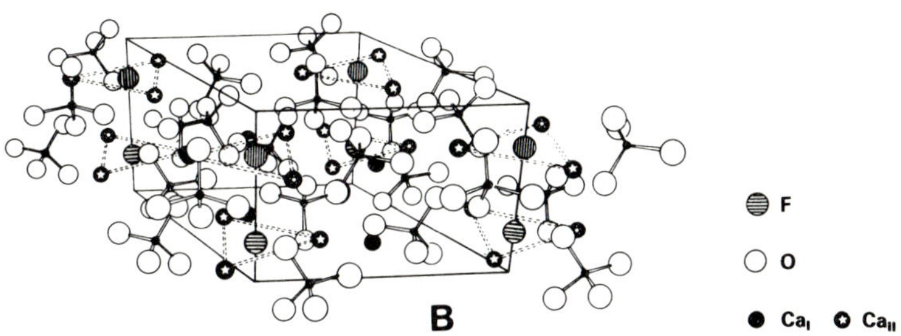

(b) Complete perspective view (from Young, 1975).

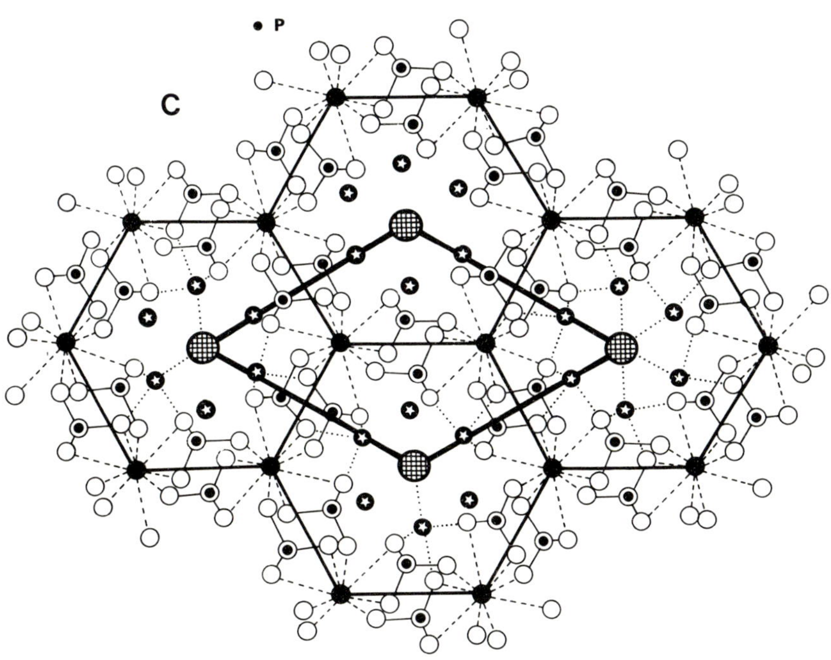

(c) Projection of a unit-cell onto plane 001.

substitution of the ions they contain or which are in their immediate vicinity. It is therefore reasonable to expect that the secondary enrichment of an apatite in uranium, by substitution of Ca^{2+} by U^{4+}, mainly affects the Ca_{II} ions of the system and not the less accessible Ca_{I} ions. Note too that Altschuler *et al.* (1958) consider that only U^{4+} can be involved in this substitution because the oxidized form U^{6+} occurs in the form of uranyl ions $(UO_2)^{2+}$, whose ionic radius (3.4 × 1.4 Å) is too large to allow penetration by way of the channels.

1.4.2. Influence of substitutions on structure

The substitutions that affect sedimentary apatites may affect the symmetry of the crystal lattice and its parameters.

As far as symmetry is concerned, it is generally accepted that the natural chlorapatites, in which Cl replaces F in the channels of the system, are monoclinic, and no longer hexagonal like fluorapatite, with a parameter *b* that is double the *a* parameter. According to Elliot, Mackie and Young (in Young, 1975), this is frequently the case with hydroxylapatite as well. Furthermore, McConnell (1974) considers that many varieties of carbonate fluorapatite are also monoclinic, or even triclinic, although these changes in hexagonal symmetry are not observable by X-ray diffraction methods.

Variations of lattice parameters associated with substitutions, on the other hand, appear to be more easily measurable, and more closely associated with specific substitutions. This is the case, for example, in the replacement of PO_4^{3-} by CO_3^{2-}, which produces a significant modification of parameter *a,* as can be seen in Fig. 5, based on the data published by Lehr *et al.* (1967).

Parameter *a* is found to decrease from about 9.370 Å to about 9.320 Å as the CO_3/PO_4 molar ratio increases from 0 to 0.3. At the same time, parameter *c* tends to increase, but the variation, which on the whole remains within the range 6.880 – 6.900 Å, is less regular and less significant.

The reduction in parameter *a* may be attributed to the shorter distance separating the O atoms in the CO_3^{2-} ions (2.11 Å as opposed to 2.54 Å in the case of PO_4^{3-}). Figure 6 (which should be compared with Fig. 3e), drawn from information provided by Bonel *et al.* (1975), summarizes the possible associated disturbances caused by this replacement near the helicoidal senary axis; they are marked with numbered squares. According to Lehr *et al.* (1967), O is replaced by F at position 1. However, this suggestion is not confirmed by Sudarnan and Young (in Young, 1975), or by Bonel *et al.* (1975), who suggest that there is in fact a vacancy that may be filled by an H_2O molecule; they believe that Ca at position 2 is replaced by Na or by another vacancy if the medium is poor in alkalies; a vacancy also appears possible at position 3, instead of an F or OH.

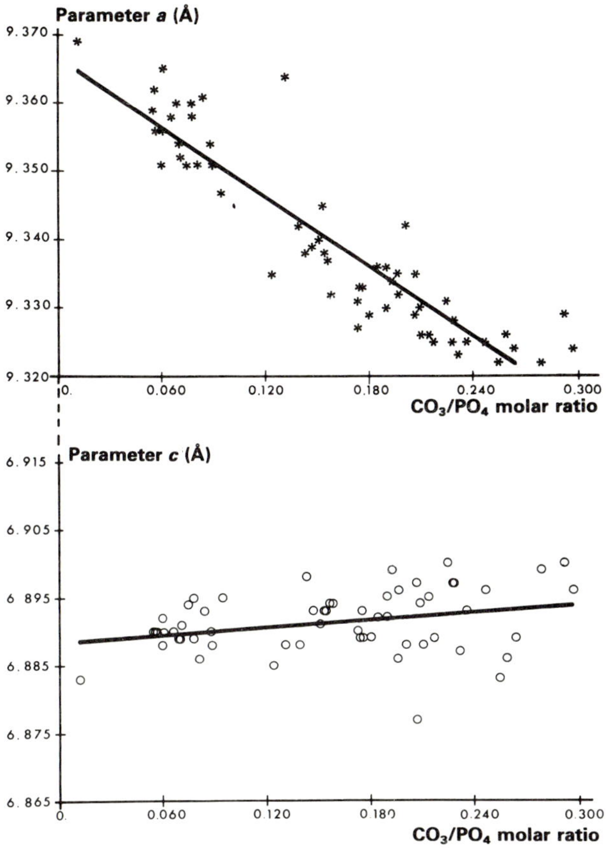

Fig. 5. *Relation between the degree of PO_4 substitution by CO_3 and parameters* a *and* c *of the lattice* (from Lehr *et al.,* 1967).

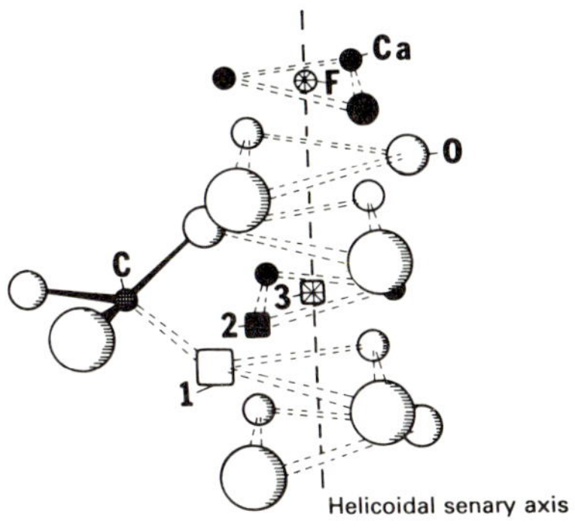

Fig. 6. *Possible disturbance of the apatite lattice by replacement of PO_4^{3-} by CO_3^{2-}* (from Bonel *et al.*, 1975).

The presence of CO_3^{2-} in this position in the lattice, although apparently confirmed by chemical analysis, has not yet been definitively proved. Brassens (1975), for example, demonstrates by thermogravimetric analysis of a Moroccan phosphate that there are two possible locations for the ion. Bonel (1972) had previously synthesized carbonate apatites in the laboratory with CO_3^{2-} ions attached to the helicoidal senary axis; these apatites (of type A) have been obtained at high temperatures, however, under conditions different from those in which natural carbonate apatites (of type B) form. The existence of natural carbonate apatites of the mixed A–B type does, nevertheless, seem possible; for example, in his study of the Bakouma deposit, Gony (1971) shows by infra-red spectroscopy that if CO_3^{2-} ions tend to replace PO_4^{3-} ions, they are also present in very small quantities in the channels, where they replace the F^- ions.

The value of parameter *a*, however, does not depend on the substitution of PO_4 by CO_3 alone. It also depends on the extent to which OH^- ions replace F^- ions. Table 4, in which analytical data for marine sedimentary phosphates (Wyoming, Montana, Idaho, Senegal, Bou Craa) are given together with those for phosphates derived from guano (Nauru Island, Christmas Island, Ocean Island) with comparable CO_3/PO_4 molar ratios, shows that the latter have both a lower F/P_2O_5 ratio and a higher value for parameter *a*. The *a* parameters for a dahllite and hydroxylapatite are, furthermore, significantly greater than those for francolite and fluorapatite, as shown in Table 5.

The reduction in parameter *a*, in conjunction with the increased substitution of PO_4 by CO_3, may to some extent be concealed by the replacement of F by

Table 4. *Relations between the weighted F/P_2O_5 ratio and the lattice parameters* (from Lehr *et al.*, 1967).

Origin	Weighted composition				CO_3/PO_4 molar ratio	Lattice parameters (Å)	
	CaO (%)	P_2O_5 (%)	F (%)	F/P_2O_5		*a*	*c*
Wyoming	47.92	32.77	3.90	0.119	0.084	9.361	6.893
Montana	52.70	37.07	3.83	0.103	0.057	9.356	6.890
Idaho	47.40	33.00	3.43	0.103	0.088	9.354	6.890
Senegal	52.42	37.46	3.83	0.102	0.071	9.352	6.891
Sahara (Bou Craa)	48.70	33.47	3.66	0.109	0.060	9.351	6.888
Ocean Island	54.73	39.82	2.72	0.068	0.055	9.379	6.884
Ocean Island	53.60	40.24	3.01	0.074	0.044	9.377	6.886
Nauru Island	53.95	38.88	2.57	0.066	0.084	9.377	6.890
Nauru Island	53.71	38.59	2.59	0.067	0.089	9.375	6.885
Christmas Island	51.90	39.35	2.23	0.056	0.078	9.386	6.880

Table 5. *Variations in lattice parameters* a *and* c *as a function of the nature of the mineral.*

Mineral	Parameters (Å)		Authors
	a	*c*	
Fluorapatite	9.37	6.88	McConnel, 1973
	9.364 ± 0.005	6.879 ± 0.005	Young, 1967
	9.367	6.884	Sudarnan *et al.*, 1972 (in Young, 1975, p. 27)
Francolite	⩽9.36	<6.89	(variable with H_2O and CO_2 present) McConnel, 1973
	9.368 ± 0.002	6.890 ± 0.002	Beatty *et al.*, 1966
	9.346 ± 0.002	6.887 ± 0.002	Brophy and Nash, 1968
Dahllite	9.436 ± 0.002	6.886 ± 0.002	Brophy and Nash, 1968
	<9.45	>6.90	(variable with H_2O and CO_2 present) McConnel, 1973
Hydroxyapatite	9.418	6.880	Sudarnan and Young, 1968 (in Young, 1975, p. 27)

OH, but if the value of the F/P_2O_5 ratio does not fall to below that of an ideal fluorapatite (0.089) then there is a good correlation between the *a* parameter and the proportion of CO_3. McClellan and Lehr (1969) have shown that the correlation is not really linear, as the effect of increasing CO_3 content on parameter *a* tends to diminish. They conclude that the substitution of PO_4 by CO_3 reaches a maximum that cannot be exceeded without destroying the lattice. Depending on the mathematical model used, the maximum number of CO_3 moles permissible in the carbonate fluorapatite formula is 1.395 or 1.722, with a minimum value of parameter *a* of 9.321 Å. This minimum value is supported by experimental data, which also show that the upper limit for the molecular CO_3 content is about 1.5. The authors mentioned above also demonstrate that parameter *c* tends to increase with increasing F content. This correlation is, however, not as good as that between parameter *a* and CO_3 content.

1.4.3. Evaluation of substitutions using crystallographic parameters

The crystallographic parameters of carbonate fluorapatites are calculated from the position of the peaks in their X-ray diffractogram. McClellan and

Lehr (1969, p. 1382) propose a number of mathematical models to enable the molar CO_3 content and – with far less certainty – the molar F content to be deduced from these parameters.

Using the data already mentioned. Gulbrandsen (1970) proposes a more direct method of calculating the percentage of CO_2 by weight from parameter *a*, using the equation:

$$y = 1014.66 - 108.253x$$

where *y* is the percentage weight of CO_2 and *x* is parameter *a* in Å; the uncertainty for the value of *y* is 0.5569. Gulbrandsen also proposes a more rapid method, dispensing with the need to calculate crystallographic parameters for determining the percentage weight of CO_2 in the apatite. As the substitution of PO_4 by CO_3 mainly affects parameter *a* and has little effect on parameter *c*, he can determine substitution directly by the difference, expressed as an angle, between the peak associated with *c* and that associated with *a* on the X-ray diffractogram. Plotting the CO_2 content as a function of the angular deviation calculated between selected pairs of peaks, he establishes a linear relationship between the two parameters. For peaks (004)–(410), he obtains the equation:

$$y = 23.6341 - 14.7361x$$

where *y* is the percentage weight of CO_2 and *x* is the angular deviation 2θ, in degrees, between the two peaks. For peaks (002)–(300)

$$y = 185 + 25.574x$$

There is an uncertainty factor of 0.5569 in the value of *y* in the first case, and of 0.5295 in the second. These relations are drawn from the diffractogram using a copper anticathode.

The pair (004)–(410), which gives higher angular values, is preferred by Gulbrandsen. If the apatite is less satisfactorily crystalline, then the (002)–(300) peaks, which are more clearly distinguished and more intense, give better results (Chaabani, 1978). In both cases, the presence of dolomite is a difficulty.

1.5. Influence of substitution of PO_4 by CO_3 on physical properties

Work carried out by Lehr *et al.* (1967) suggests that there is a direct relation between the size of the apatite crystals, which varies between 400 and 1200 Å in the sedimentary ores studied, and the value of parameter *a*, and an inverse relation between size and solubility of apatite in citrates or acids. In other

Table 6. *Influence of PO_4 replacement by CO_3 on the birefringence characteristics of fluorapatite* (from McConnel, 1973).

CO_2 (%)	F (%)	*a* (Å)	Birefringence index
4.39	4.59	9.334	0.0125
4.22	4.06	9.342	0.0098
3.18	3.83	9.349	0.0082
2.57	3.94	9.350	0.0070
2.01	3.83	9.352	0.0062
1.32	3.71	9.360	0.0055
1.26	3.78	9.361	0.0051
—	3.78	9.371	0.0034

words, the solubility of phosphate minerals increases with increasing substitution of PO_4 by CO_3, while crystal size decreases.

This substitution also leads to a decrease in the apatite refractive index from 1.64 to 1.60, as the CO_3/PO_4 molar ratio increases from 0 to 0.3; the birefringence index increases at the same time, as can be seen in Table 6, based on Carstrom (in McConnell, 1973, p. 44). This effect of carbonate substitution is important in petrographic studies of thin sections, which may enable a good approximation to be obtained for the degree of substitution, but can also lead to confusion between carbonate apatite and chalcedony, particularly if only a superficial examination is made.

2. OTHER PHOSPHATE MINERALS ASSOCIATED WITH SEDIMENTARY DEPOSITS

The weathering of calcium phosphate deposits and guano, and diagenesis occurring in certain lacustrine or fluviatile sediments, are the main factors responsible for the development of non-apatitic phosphate minerals in sedimentary formations.

Lateritic alteration of pre-existing deposits of calcium phosphate leads primarily to aluminium phosphate minerals, which may be very common, as in Senegal, on the Benin–Burkina Faso (formerly Upper Volta)–Niger border, in Brazil, in Venezuela, and on Christmas Island. These minerals have been studied in particular by Capdecomme (1952, 1957) in the Thiès region of Senegal; he established the presence of the following species:

(a) Millisite: $(Na\text{-}K)CaAl_6(PO_4)_4(OH)_9 3H_2O$, and pallite, iron bearing varieties of millisite with a low sodium content.

(b) Crandallite: $Ca_2Al_6(PO_4)_4(OH)_{10} 2H_2O$.

(c) Augelite: $Al_2(PO_4)(OH)_3$.

(d) Wavellite: $Al_3(PO_4)_2(OH)_3 5H_2O$.

(e) Turquoise: $CuAl_6(PO_4)_4(OH)_8 4H_2O$.

The first three minerals are the most common; wavellite and, particularly, turquoise are rare. The chemical composition of these minerals varies widely, differeing from the ideal formula because they contain F, and significant proportions of Sr, Ba or Fe. An analysis of Senegalese crandallite, for example, allowed Capdecomme and Orliac (1967) to establish the following structural formula:

$$Ca_{0.81}Ba_{0.15}Sr_{1.06}Al_{6.12}(PO_4)_4(OH)_{9.30}F_{1.10} 3.15H_2O$$

which is intermediate between that of *crandallite, goyazite* $[Sr_2Al_6(PO_4)_4(OH)_{10} 2H_2O]$, a mineral reported by the United Nations mission from the weathered zone of the Infracambrian phosphates of northern Benin, and *gorceixite,* which is essentially barytic.

Vivianite, $Fe_3(PO_4)_2 8H_2O$, may also appear in a weathering profile of this kind (e.g. at Taïba, Senegal), but it is mainly found in lacustrine or alluvial deposits, in a reducing environment, where it may be associated with reddingite $[Mn_3(PO_4)3H_2O]$. According to Nriagu and Dell (1974), in a more oxidizing environment these minerals tend to be associated with tinticite $[Fe_3(PO_4)_2(OH)_3 3H_2O]$, lipscombite $[Fe_3(PO_4)_2(OH)_2]$, cacoxenite $[Fe_4(PO_3)_3 12H_2O]$, rockbridgite $[Fe_5(PO_4)_3(OH)_5]$, strengite $[FePO_4 2H_2O]$, . . . or, in the presence of calcium, with foucherite $[CaFe_4(OH)_8(PO_4)7H_2O]$, mitridatite $[CaFe_2(PO_4)_2(OH)_2 8H_2O]$, or even anapaite $[Ca_2Fe(PO_4)_3 4H_2O]$ at pH values less than 7.

Fresh guano contains about 4% P_2O_5 (McKelvey, 1973). They then lose their most volatile and most soluble constituents; the magnesium and ammonium phosphates they contain are concentrated and react with the rocks on which they accumulate.

On calcareous rocks, calcium phosphates form, such as monetite ($CaHPO_4$) or whitlockite $[Ca_3(PO_4)_2]$ and, above all, apatitic minerals relatively poor in fluorine; the fluorine content may increase later.

On siliceous rocks, aluminocalcic phosphates form, such as crandallite, or phosphates of iron and aluminium which tend to be classed together under the name of 'barrandite', a solid solution of minerals whose extreme forms are strengite and variscite ($AlPO_42H_2O$). The presence of montgomeryite [$Ca_4Mg(H_2O)12Al_4(OH)_4(PO_4)6$], associated with crandallite, has been revealed in a cave deposit originating from bat guano (Peneaud, 1978).

Other phosphate minerals have been reported in sedimentary deposits (Fischer and Jerome, 1973). Among them are monazite, a rare earth phosphate, which is only found in Ordovician rocks, in particular, in Brittany, northern France (Donnot *et al.*, 1973), and autunite [$Ca(UO_2)_2(PO_4)_210$-$12H_2O$], or meta-autunite [$Ca(UO_2)_2(PO_4)_22$-$6H_2O$], secondary minerals of certain deposits rich in uranium, such as at Bakouma, in the Central African Republic (Gony, 1971).

Chapter 3

Petrography of sedimentary phosphates

The wide variety of petrographic characteristics of sedimentary phosphates makes it necessary to use a specific terminology to describe and compare the rocks in which they are found. We shall therefore begin by proposing a classification system.

1. PETROGRAPHIC CLASSIFICATION

The terminology proposed for describing and classifying phosphate rocks has been drawn up with three aims in mind: to limit the wide range of petrographic terms in current usage, to improve communication between geologists of different nationalities, and to facilitate cooperation with other specialists (mining and ore-dressing engineers, etc.). The terminology proposed is consequently as descriptive as possible and no more 'generic' than is strictly necessary. An effort has also been made to use familiar or etymologically simple terms easily adaptable to other languages.

1.1. Definitions

1.1.1. Phosphate rocks

There is still no general agreement on the petrographic designation of sedimentary rocks with a predominant phosphate content. 'Phosphate', often used in this sense in France, is not a good petrographic term, as it belongs strictly to the chemical field (where it is used for salts of phosphoric acid). It should not be used. 'Phosphorite' seems more suitable at first sight. Being common to a number of languages, this term is already widely used throughout the world for sedimentary rocks rich in phosphate. Its use in this sense is also becoming current in France, where it was traditionally reserved for the secondary concentrations of phosphates in karstic pockets (e.g. the Quercy phosphorites).

Unfortunately, the many users of this term are by no means in full agreement about its exact meaning. Some authorities (Bushinski, 1964; *Soviet Geological Dictionary,* 1973; Pettijohn, 1957; Swanson, 1970; etc.) imply that in a phosphorite, the phosphate content must be the predominant fraction. For Riggs (1979a), on the other hand, a phosphorite has only to contain a minimum of 10% by volume of phosphate grains, while for Gimmelfarb *et al.* (1959), only 5% P_2O_5 content is necessary. With less precision, Altschuler *et al,* (1964) define a phosphorite as "a rock or specimen containing substantial amounts of sedimentary apatite". The latest 1980 edition of the *Glossary of Geology* published by the American Geological Institute has a similar definition: "a sedimentary rock with a high enough content of phosphate minerals to be of economic interest."

There is no greater measure of agreement on the mineralogical side. For example, according to McConnel, the phosphate fraction must consist of calcium phosphates (in Pettijohn, 1957) or even made up only of francolite (McConnel, 1973). However, Cook (1976) and Riggs (1979a) are fully prepared to accept aluminium phosphate minerals within the definition of phosphorites.

The previous edition of this work (Slansky, 1980) proposed that the term phosphorite should be used for sedimentary rocks consisting mainly of phosphate minerals, which might be phosphates of calcium, aluminium or iron. It has been found, however, that it would be better to give this term a wider meaning, extending it, provisionally at least, to cover any rock having an appreciable phosphate fraction, without attempting to include it in any rigorous petrographic system until general agreement on its definition can be reached.

'Phosphatite' is therefore the only term used in this second edition for a sedimentary rock consisting mainly of phosphate minerals. This etymologically simple term may be compared with 'argillite', used by Millot (1949, 1963) for rocks consisting mainly of one or more clay minerals.

'Phosphatite', moreover, is not a new word, but appears amongst the petrographic terms in the *Glossary of Geology* [(1)]. It has the advantage over 'phosphorite' of not yet being widely used, so that its rigorous use is not constrained by established habits.

---

[1] In its 1974 edition, the *Glossary of Geology* defines a phosphatite as "a sedimentary phosphatic rock composed of the mineral apatite in its various forms, e.g. phosphate rock". In the subsequent 1980 edition, however, the term is treated simply as synonymous with phosphorite.

The term 'phosphatite' without a prefix will be used for the most common rocks, with a fraction made up of calcium phosphates, i.e. mainly apatite. *A phosphatite is therefore a sedimentary rock composed mainly of minerals of the apatite family.* A prefix provides a means of naming rocks with other phosphate minerals, aluminophosphatite, perhaps ferriphosphatite, or even uranophosphatite.

Furthermore, when a sedimentary rock contains 50% phosphate minerals, its P_2O_5 content depends on the nature of the phosphate minerals present. If fluorapatite predominates, the P_2O_5 content can exceed 21%; but in a carbonate fluorapatite this content may be less than 17% if substitution of PO_4 by CO_3 is important. In a rock with 50% crandallite or augelite, the P_2O_5 content is of the order of 20%, while it is greater than 22% in variscite, but if the predominant mineral is vivianite or goyazite, the P_2O_5 content may be below 16%.

In view of this variation, a P_2O_5 content of 18% has been arbitrarily selected as the borderline between rocks regarded as having over 50% phosphate minerals and those regarded as containing less than 50%; a P_2O_5 content of 36% is a good general figure for the average carbonate apatite (francolite), in which one of the six PO_4 ions has been replaced by CO_3.

Under these conditions, the terms 'phosphatite' and 'aluminophosphatite' can only be used for rocks having a P_2O_5 content of 18% or higher. If the P_2O_5 content is less than 18%, the primary petrographic term to be used is that for the non-phosphatic fraction, with 'phosphatic' as a qualifier – e.g. phosphatic sandstone, clay, chalk, etc. A more rigorous terminology (Scolari and Lille, 1973) uses terms such as quartzarenite with argillo-phosphatic cement, phosphatic biocalcilutite, etc.

From a practical point of view, therefore, the term 'phosphorite' will be applied to rocks with an uncertain phosphate content or to a group of rocks in which this content varies. 'Phosphatite' and 'phosphatic' will be terms used only for rocks whose phosphate content is known, or in which it can at least be determined, whether or not the phosphate content plays a dominant role.

In a phosphorite or a phosphatite, the non-phosphatic fraction is traditionally termed the 'endogangue' (Visse, 1953) when it is inside the grains, nodules or other phosphate particles and the 'exogangue' when it is outside them. The siliceous or carbonate coating that more or less encloses certain grains may be called the 'epigangue'.

1.1.2. Phosphate particles

In a descriptive terminology, the phosphate particles are primarily defined by their size and shape. The smallest particles are grains whose size is smaller than that of granules, i.e. 2 mm in diameter. Above 4 mm, these particles are

Table 7. *Subdivisions of the rudite class.*

Grain size	Subdivision of rudite
2 mm	
	granule
4 mm	
	gravel
16 mm	
	pebble
64 mm	
	cobble
256 mm	
	boulder

'gravels', then pebbles, cobbles and boulders, as shown in Table 7. Their shape may be rounded, half-rounded, sub-angular, angular or flat.

In mined phosphate ores, rounded, half-rounded or more or less ovoid grains are generally the most common and were often called 'ooliths'. Their general lack of concentric structures, however, has gradually led to these grains being called 'pseudo-ooliths', and now, most commonly, 'pellets'.

More or less strict definitions of the term 'pellet' are given by different authors. In general use it has the widest possible meaning. According to Pettijohn (1957, pp. 94 – 95), these grains, whose origin is not necessarily faecal and which are generally rounded to half-rounded, may also have less regular shapes; being cryptocrystalline or microcrystalline, they have no radial or concentric structure. Pellets may also contain various organic or mineral inclusions: formaminiferal or radiolarian tests, frustules of diatoms, undifferentiated organic matter, fragments of quartz, feldspar, limestone, dolomite rhombohedra, etc. The size of phosphate pellets may in principle vary from 10 μm to 2 mm, but in practice it is usually between 50 μm and 1 mm.

Ooliths differ from rounded pellets in that they have a clear internal concentric structure. Pisolites of comparable structure measure over 2 mm in diameter.

Bioclasts, intraclasts, lithoclasts, coproliths and phosphate nodules are also well known particles of phosphatites. Since they are defined by their origin, however, these terms can only be used when the origin is clearly evident.

Bioclasts are tests or skeletons of organisms (or fragments of them) that have undergone 'transport', even if not very far, and have been deposited by gravity

(Scolari and Lille, 1973). To be considered as bioclasts among the phosphate particles, these fossils have to be clearly phosphatic either in origin or by secondary replacement, and still be identifiable externally. So, a grain with a Foraminiferal test, totally concealed by the phosphate layer enclosing it, is not a phosphatic bioclast but a pellet, or possibly an oolith, with a bioclast.

Identifiable internal moulds (of pteropods, ammonites, etc.) are, for convenience, grouped here as bioclasts.

Intraclasts are reworked fragments of a contemporary sediment of the same depositional basin, still comparatively unconsolidated.

Lithoclasts are reworked fragments of a pre-existing phosphatite, which have already been lithified and deposited mechanically after being transported over a variable distance.

Coproliths are fossilized organic excreta enriched with phosphate and lithified in the course of diagenesis. In English-speaking countries this term (as, for example, in Pettijohn (1957), is often used only for the fossilized excrement of vertebrates, with sizes varying from a few millimetres to over ten centimetres, the term 'faecal pellets' being employed for the fossilized excrement of invertebrates, often less than a millimetre long. This distinction no longer appears necessary in the case of phosphate rocks, in which the faecal origin of the pellets remains uncertain.

Phosphate *nodules* are elements measuring anything between several millimetres and several tens of centimetres, found in various shapes – spherical, lenticular, flattened parallel to the sedimentary bed, etc. – and formed by chemical or biochemical concretion around one or more centres, most frequently of organic origin.

1.1.3. Crystalline textures

Sedimentary apatites are frequently *cryptocrystalline*, i.e. the crystals are too small to be visible under the microscope in thin section; in polarized light they appear isotropic. When the size of the crystals reaches or exceeds a few μm, however, their birefringence becomes very clear; the texture is then *microcrystalline*. An upper limit to the size of the crystals can be set at 63 μm, based on suggestions made by Scolari and Lille (1973) for carbonates. Above this size and up to 2 mm the texture is *macrocrystalline* and then *megacrystalline*, but these textures do not appear in sedimentary apatites unless they have already undergone considerable metamorphism.

This terminology can naturally also be used for the carbonate siliceous or gypseous gangue of phosphatites.

1.2. Phosphatite nomenclature (P_2O_5 > 18%)

The classification proposed is essentially intended to be descriptive; it is comparable with that of Scolari and Lille (1973) for carbonate rocks, and also draws, in a simplified form, on the work of Folk (1959). Following Folk's recommendation for carbonate sediments, the phosphate particles are only taken into account if they are more than 10 μm in diameter. The name of the rock depends on their presence and their size, as soon as they represent at least 10% of the rock in question.

1st case: The phosphatite contains less than 10% phosphate particles.

As in limestone micrite, the grain size of the phosphate fraction is, in this case, generally less than 10 μm. A rock of this type could be called 'collophanite', i.e. a rock consisting mainly of collophane, which is a cryptocrystalline or incompletely crystallized form of sedimentary apatites. The term 'collophanite', however, has already been used with various different meanings and does not fit satisfactorily into modern systems of terminology; 'microsphatite' seems preferable, or possibly 'microsphorite' if the mineralogical nature of the phosphate is uncertain.

The microsphatite may be homogeneous, laminated, of stromatolitic or concretionary structure, etc. If it contains more than 10% non-phosphatic material, the microsphatite may be siliceous, argillaceous, glauconitic, calcareous or, if greater precision is desired, quartzarenitic, biocalcarenitic, etc. With less than 10% foreign matter, the microsphatite may be characterized by rare siliceous nodules, dispersed glauconite grains, fine silty beds, etc.

2nd case: The phosphatite contains more than 10% phosphate particles.

This phosphatite is by far the most common. The name of the rock then depends on the size of the most numerous grains: if the size of these grains is below 63 μm, it is a phosphalutite; if it is between 63 μm and 2 mm, it is a phospharenite; if it is above 2 mm, it is a phospharudite.

Terms such as 'phospharenolutite' or 'phospharenorudite' may be used for the distribution of grains between the two grain-size categories.

If still greater precision is needed, the nature of the most abundant particle may be introduced into the name of the rock by the use of the prefixes bio, litho, intra, oo and pel, derived respectively from bioclast, lithoclast, intraclast, oolith and pellet. Examples are: lithophospharenite, intraphospharudite, oophospharenite, pelphosphalutite, biophospharudite, biopelphospharenite, etc. As ooliths belong by definition to the 'phospharenite' class, the term 'oosphatite' could be used in place of 'oophospharenite', but as the phosphatites with predominant true oolilths are comparatively rare, this modification of terminology seems unnecessary.

As there is no upper limit to the size of members of the rudite class, subdivisions by grain size have been introduced to make the terminology more precise for the phospharudites. The subdivisions proposed are as follows: granule (2 – 4 mm), gravel (4 – 16 mm), pebble (16 – 64 mm), cobble (64 – 256 mm), boulder (> 256 mm), as shown in Table 7. The size of the rudites may be specified by the use of adjectives: e.g. granular phospharudite, pebbly biophospharudite, cobbly lithophospharudite (Slansky, 1979a). In the case of a distinctly characterized bed with bone debris, the internationally accepted term 'bone bed' may be added in parentheses after the name of the rock: e.g. sub-angular pebbly and cobbly biophospharudite (bone bed).

Phosphatites may also be characterized by a second, less common phosphate fraction, by their endogangue and their exogangue.

This second phosphate fraction, when significant, may be indicated by an adjective: e.g. microsphatitic pelphospharenite, arenitic phosphalutite, arenolutitic biophospharudite. If this subordinate fraction is of the same class as the rudites, the use of 'pebbly', 'cobbly', etc. is preferred to the insufficiently precise adjective 'ruditic'; the shape of the pebbles, cobbles, etc. can then be specified, e.g. 'phospharenite with rounded gravel'. When the nature of these large subordinate particles is characteristic enough, it may be useful to give the size in centimetres, within parentheses: e.g. pelphospharenite with internal gastropod moulds (5 –10 cm), with 'fish teeth' and bone debris (2–5 cm), etc. This method of characterizing subordinate phosphate material is applicable to all phosphatites, and therefore also to microsphatites.

No specific terminology is required to indicate the nature of the endogangue, though it is possibly a good idea to give an indication of its relative extent, e.g.:

(a) Pelphospharenite with rare endogangue of fine quartz.

(b) Biophospharudite with sub-rounded small pebbles and abundant calcite and glauconite endogangue.

(c) Phospharenite with widespread endogangue of fine pyrite and diffuse organic matter.

The exogangue, if present, assumes an importance in phosphatite terminology comparable with that of the size of the phosphate grains. It is characterized by its mineral composition, the size of its grains or crystals and its relative abundance in the rock, e.g.:

(a) Biopelphospharenite with less than 10% exogangue.

(b) Pelphosphalutite with 10% siliceous microcrystalline exogangue.

(c) Biophospharenite with 30% dolomicrite exogangue.

(d) Phospharenite with 20% sparry calcite (or sparite) exogangue.

(e) Biophospharenite with 10% quartzsiltite exogangue.

(f) Lithophospharudite with 30% argillo-glauconitic quartzarenite exogangue.

Finally, the name of a phosphatite may be completed by a reference to the texture, hardness and colour of the rock. Dunham's terminology (Dunham, 1962) – grainstone, packestone, wackestone – may be used for texture, or Pirlet's (Pirlet, 1965) – 'texture jointive', 'texture empâtée'. The texture may frequently be deduced, however, directly from a good description of the phosphate fraction and the exogangue. Hardness is of importance for the mine operator but difficult to define in a simple petrographic terminology; it is sufficient, at this stage, to use just 'consolidated' or 'unconsolidated' if these properties are clearly shown. Colour frequently depends on the conditions under which the specimen was obtained.

In summary, the proposed phosphatite terminology provides information on the nature and grain-size category of the main phosphate fraction, the grain-size category of the second phosphate fraction, the nature and extent of the endogangue, and the nature, grain-size category and relative abundance of the exogangue.

Such a terminology is obviously too cumbersome to be constantly used in its entirety. In general discussion, the simplest possible term will be used, with an attempt to use the grain-size category of the main phosphate fraction as the name of the rock; for example, 'phospharenite' is often an adequate name on its own for the North African or Far Eastern ores. In applied geology, it is desirable for a certain P_2O_5 percentage range to be added, e.g. phospharenite with 26 – 30% P_2O_5.

If a greater degree of precision is required, the nature of the exogangue will be added first, followed perhaps by its relative abundance. The grain-size category of the second phosphate fraction and the characteristics of the endogangue will not be given except where there is a particular need for such supplementary information.

1.3. Classification of phosphatites

The terminology defined earlier enables us to classify phosphatites by their physical characteristics, the two basic parameters being the grain size of the main phosphate fraction and the nature of the exogangue (or the non-phosphatic fraction, in the case of the microphatites).

The nature of the exogangue may vary considerably; it can nevertheless be subdivided into five main categories according to its predominant petrographic fraction:

(a) Clastic siliceous exogangue, in which grains of quartz measuring over 10 μm predominate.

(b) Cherty exogangue, in which silica with a grain size below 10 μm predominates.

(c) Clayey or marly exogangue. This category includes glauconitic exogangue.

(d) Carbonate exogangue.

(e) Exogangue not included in the preceding categories and whose predominant fraction may, for example, be gypseous, ferruginous or contain pyroclastic silicates.

Table 8. *Major phosphatite classes: for example box* A *represents a cherty phospharenite.*

Main phosphate fraction	Exogangue				
	Clastic siliceous (>10 μm)	Cherty (<10 μm)	Clayey/ marly	Carbonate	Other
Microsphatite					
Phosphalutite					
Phospharenite		A			
Phospharudite					

Twenty main classes of phosphatite may be defined in this way, as shown in Table 8. In the deposits worked at present, the phospharenites predominate, with an exogangue that is mainly clastic siliceous, clayey-marly or carbonate.

Where necessary, further subdivisions can be made, as discussed earlier.

1.4. Other phosphorites

Among the sedimentary rocks with an abundant but not apatitic phosphate fraction, the aluminium phosphorites, or aluminophosphorites, are undoubtedly the most common. Applying our definitions, aluminophosphatites are aluminophosphorites whose P_2O_5 content is at least 18%. These rocks are chiefly produced by the weathering of argillaceous calcium phosphate deposits.

There have so far not been many petrographic studies of aluminophosphatites, but the work that has been carried out, particularly the recent study by Flicoteaux (1982) on the genesis of aluminium phosphates of western Senegal, nevertheless allows a terminology based on the macroscopic appearance of the associated non-phosphatic fraction, and the nature of phosphate minerals.

In macroscopic terms, the aluminophosphatites are mainly bedded, laminated, striped, in plates, brecciated, nodular, vacuolar or massive. Their non-phosphatic fraction, if any, is primarily siliceous, argillaceous or ferruginous. Crandallite, millisite, wavellite and augelite are the main phosphate minerals they contain.

Examples of types of aluminium phosphate rocks are therefore bedded aluminophosphatites with crandallite, or brecciated siliceous aluminophosphatites with crandallite and millisite, or vacuolar aluminophosphatites with augelite, etc.

2. FORMS OF THE PHOSPHATE FRACTION OF PHOSPHATITES

Cayeux (1950, p. 919), writing on the subject of calcium phosphate grains, points out that the grain unit varies so widely in its constitution and its history that it may be regarded, without any exaggeration, as forming a little world of its own. The wide diversity of form of the phosphate fraction has, if anything, been shown to be even greater by modern means of observation and analysis. It is therefore, at present, very difficult to give details of all the forms in which the phosphate fraction of phosphatites may occur, and the sole aim of the following pages is to give some ideas on the subject.

2.1. The microsphatites

At the stratum level, microsphatites are not frequent and often appear to be of secondary origin. Plate 1 illustrates four examples from Gabon, Greece, India and Brazil.

In Gabon, a microsphatite of Upper Cretaceous age outcrops in large blocks, variously distributed in the bed of Benghe River, in the southern part of the coastal sedimentary basin. Underneath the surface patina the rock is very white, with a very fine homogeneous grain and areas of fine scattered quartz grains; in the absence of silica, its P_2O_5 content may exceed 38%. The colour of the rock and the context indicate extensive weathering. Photographs 1 and 2 reveal the microcrystalline nature of the apatite, which generally corrodes the grains of quartz present.

In Epirus, Greece, a microsposphatite forms part of the Middle Jurassic sequence, varying in colour from very dark grey to black. It may occur in lenses, but it is found mainly in fracture fillings in the Mitsikeli Massif. The zones of calcite that persist within the microcrystalline apatite (Photograph 3) indicate a secondary origin for the rock, by epigenesis of pre-existing limestone. Further evidence of secondary origin is provided by an abnormally high uranium content.

In India, in the Mussoorie region, to the north-north-east of Delhi, the phosphate sequence previously attributed to the Jurassic but recently allocated to the Permian or even to an earlier Palaeozoic period, includes phosphate beds with a microsphatite facies amongst cherts and phospharenites. Examination of a thin section (Photograph 4) shows, however, that this microsphatite derives from a pre-existing phospharenite whose grains have been welded by a microcrystalline apatite. In Minas Gerais, Brazil, certain phosphate beds of homogeneous appearance in the Infracambrian Bambui Basin show similar features, although it can be even harder to find a trace of the original grains because the rocks have been subjected to slight metamorphism.

The presence of primary microsphatite in strata is, however, accepted by several authors: e.g. Riggs (1979b) for Florida, Howard and Cooney for the Cambrian in Australia (in Cook, 1976).

2.2. Particles

In the vast majority of cases, the phosphate occurs in deposits in the form of particles of the size of arenites. Most ores are therefore phospharenites or possibly rocks with phosphate particles of the size of arenites, such as phosphatic siliceous sands, or quartz-arenites, with phosphate pellets in the upper layers of the Mio-Pliocene deposits in Florida.

The grains of phospharenites in situ do not generally reach the size of lutites. Among the ores studied at the B.R.G.M., the specimens richest in grains of lutite size are from the Cambrian deposits in Queensland, Australia, and primarily from the Infracambrian of Burkina Faso which can reach 35 μm; in both cases, however, they occur in subordinate amounts in phospharenites.

Plate 1

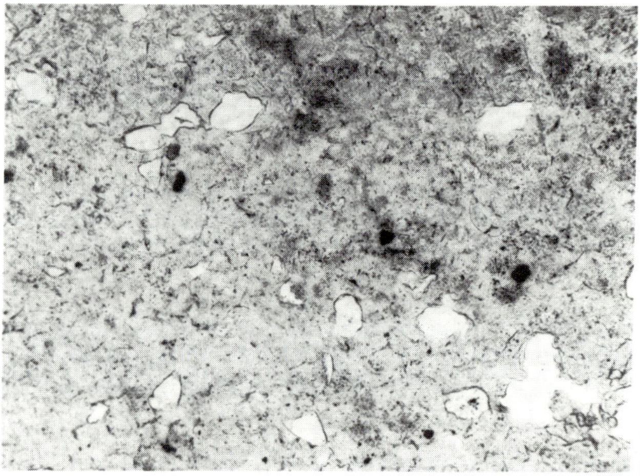

1 DL — 0.1 mm

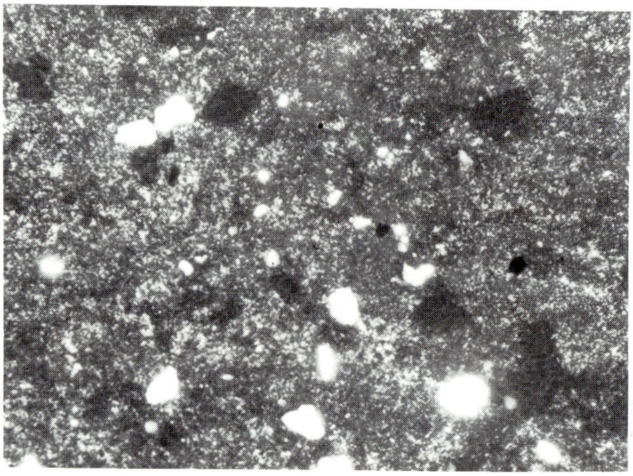

2 CP — 0.1 mm

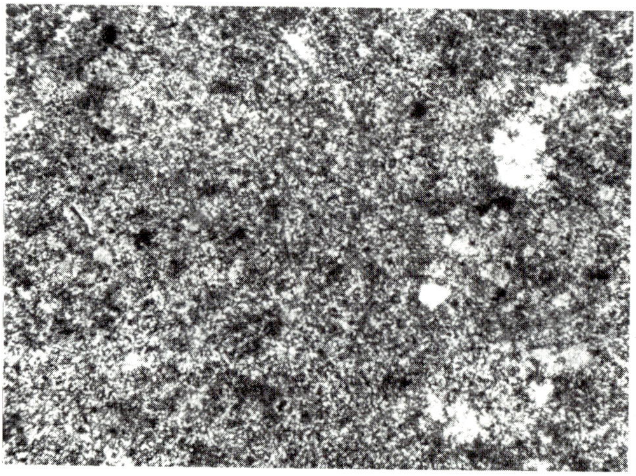

3 CP — 0.1 mm

Plate 1
Continued

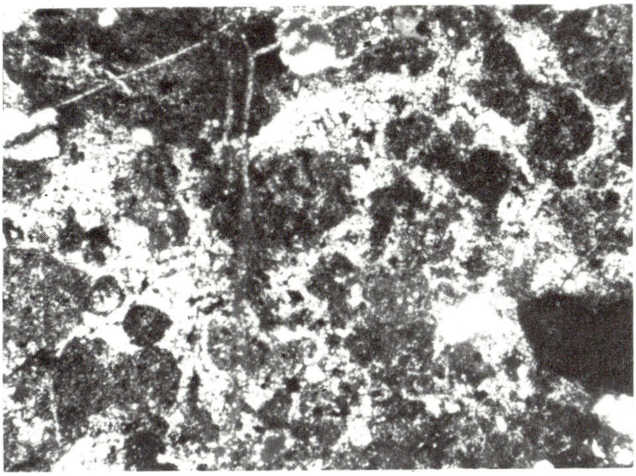

4 CP — 0.1 mm

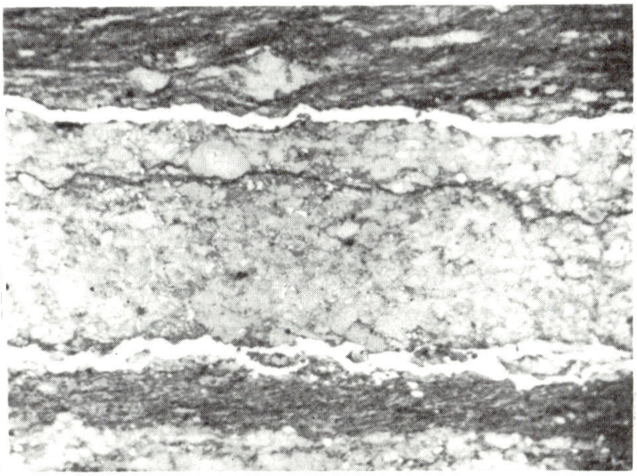

5 DL — 1 mm

DL = Ordinary light **CP = Crossed polarized**

Photomicrograph of selected phosphatites

1 and **2**: Senonian microsphatite, from Gabon. Microcrystalline apatite with small, scattered, grains of quartz.

3: Middle Jurassic microsphatite from Epirus, Greece. The apatite is microcrystalline. The white areas correspond to traces of calcite that have resisted epigenetic alteration.

4: Palaeozoic microsphatite from Mussoorie, India, containing cryptocrystalline pellets held together by microcrystalline apatite.

5: Bed of Infracambrian microsphatite (lighter area) from Minas Gerais, Brazil, composed of pellets in an apatite current.

Phospharudites are less rare than phosphalutites, though they are still not particularly common and do not generally form thick beds, apart from a few special cases: 'bone beds', local accumulations of 'coproliths', etc. As a subordinate fraction, phosphate particles over 2 mm in length are more common: 'pebbles' of Florida ore, nodules of the Albian succession from the Paris Basin or of the Dinantian in the Pyrenees and the Montagne Noire, phosphate nodules from the present sea floor, etc.

2.2.1. Particles of arenite size

Phospharenite grains, whose outward appearance can be seen in Plate 2, are generally rounded, i.e. they form pellets. There are, however, some more or less angular or flattened grains that can generally be classified as intraclasts or bioclasts (Plate 3). The pellets occur in widely varying forms, owing to the differences in the texture of their phosphate fraction and the nature of their endogangue; Plates 4, 5 and 7 give an idea of how broad this variety is.

Certain pellets are very homogeneous in appearance, with practically no endogangue; grains of this type are frequently contained in sub-outcrop strata. Although entirely apatitic, other pellets have very clear organic inclusions: debris of diatomaceous phytoplankton, Radiolaria, Foraminifera, all epigenized with phosphate; in the last two cases the fossil is generally well centred in the pellet, which has obviously formed around it.

A Foraminiferal inclusion may give rise to different varieties of pellet, according to whether or not the test has been replaced, and whether the apatite only fills the chambers or also forms one or more external layers (Plate 5). Photograph 22 illustrates the complex case of a Foraminiferal with a calcite test surrounded by phosphate, whose chambers are partly filled with apatite and partly with glauconite, with disseminations or clusters of pyrite.

Glauconite is rare in ores of greatest economic value, but frequent in lower quality phosphate strata, where it takes a number of mineralogical forms ranging from iron-bearing smectites to glauconite proper. In general, phosphate and glauconite are found in separate grains; studies of phosphate specimens of various origins (Albian from the Paris Basin, Palaeocene from Iran and Benin, Lower Eocene from Tunisia) carried out by the B.R.G.M. have, nevertheless, shown that there are commonly mixed grains of apatite and glauconite. Plate 6 illustrates this kind of close association. These two green phosphate grains are from the Albian layers in the Paris Basin. The results of spot analysis on these grains with the Castaing electron microprobe, i.e. on surfaces of the order of 1 μm^2 each (by arbitrary convention, the Fe content is expressed as the total of Fe_2O_3), shows how closely these minerals mingle, as is also made clear by observation with an transmission electronic microscope (by J.L. Boulmier) on green grains from the Palaeocene deposits in Benin and Iran: the apatite-glauconite crystals have dimensions of the order of half a micron, and are associated with Fe_2O_3 crystals of the same size.

Plate 2

6 1 mm

7 1 mm

8 1 mm

9 1 mm

Plate 2 *Continued*

10 1 mm

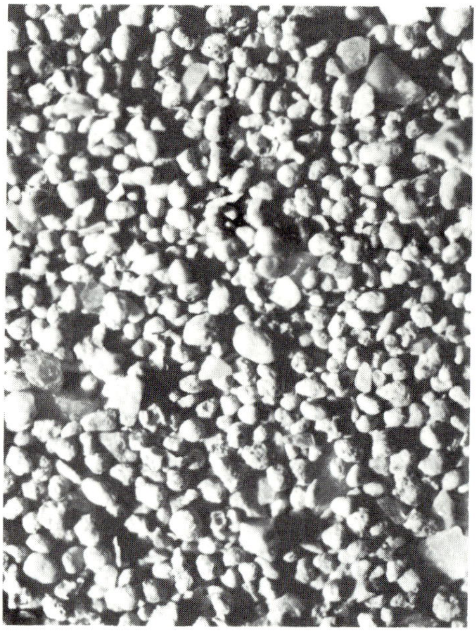

11 1 mm

6: Egyptian Senonian (Nile Valley). **7**: Saudi Arabian Senonian (West Thanyat). **8**: Moroccan Palaeocene (Bou Craa). **9**: Middle Eocene in Benin (Nigeria). **10**. Middle Eocene in Senegal (Taïba). **11**: Californian Miocene (Indian Creek).

Sassi (1974) observed similar close relationships between glauconite and apatite in phosphate grains from Tunisian deposits. He points to the possible presence of clay minerals (indicated by silicon and aluminium traces found using the electron microscope, but not specifically identified) in fine concentric layers within the pellets, which then show a clear oolithic texture. Other silicates, such as clinoptilolite, have been observed in inclusions in the pellets of Tunisian phospharenites.

Quartz may be frequent in the endogangue of phosphate pellets, at least locally (Infracambrian in Niger, Permian in the Rocky Mountains, Miocene in California). A pellet may contain several grains of quartz but generally only one, around which the apatite is distributed in one or more layers. This central nucleus is sometimes but less frequently formed by a grain of feldspar (Photograph 31), a limestone fragment, a dolomite rhombohedron, etc.

Plate 3

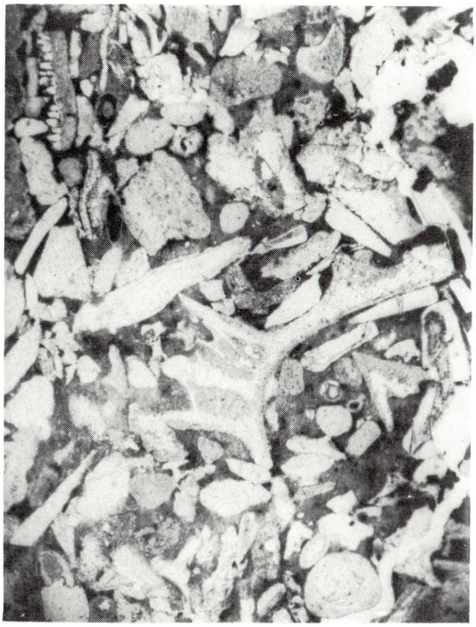

12a OL 1 mm

12b CP 1 mm

13 OL 1 mm

14 CP 1 mm

Plate 3 *Continued*

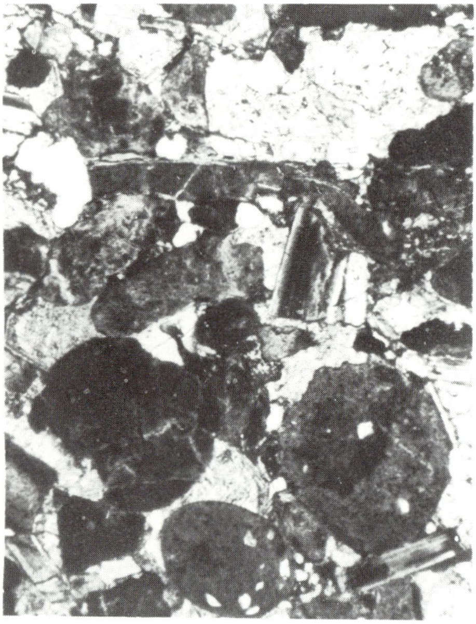

15a OL 1 mm

15b CP 1 mm

OL = Ordinary light CP = Crossed polarized light

12a and **12b**: Biopelphospharenite with bone debris, from the Egyptian Senonian. **13** and **14**: Biophospharenite (brachiopods) with a quartzarenite exogangue from the Permian of the Rocky Mountains, U.S.A. **15a** and **15b**: Intrabiophospharenite with a calcareous sandstone exogangue from the Libyan Triassic.

Finally, organic matter is always found in the endogangue of pellets that have not undergone too intensive secondary oxidation. It can appear in many different ways, in the centre, around the outside or in concentric layers; Plate 7 shows a few examples. When large quantities of organic matter are present, fine pyrite crystals may be associated with it in the endogangue (as in the Abu Tartur deposit in Egypt, the Mussoorie deposit in India, etc.). Recent studies by Belayouni (1981, 1984), whose results are confirmed by Amit and Bein (1982), Rachidi (1983) and Samb (1983), show that this organic matter is of marine planktonic origin and made up essentially of humic acids.

The outer edge of the pellets is quite often covered by a cortex consisting of a layer of hyaline apatite (in fibrous or microcrystalline form) of variable thickness, usually not exceeding a few microns. In certain cases, a single outer layer encloses a number of pellets (forming a compound pellet); in others, a number

Plate 4

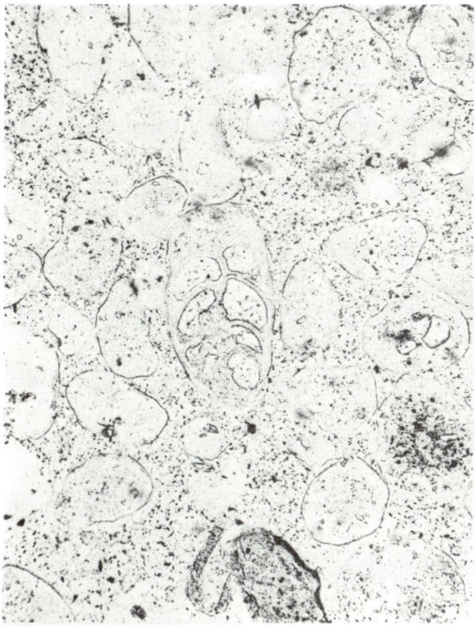

16a OL 0.1 mm

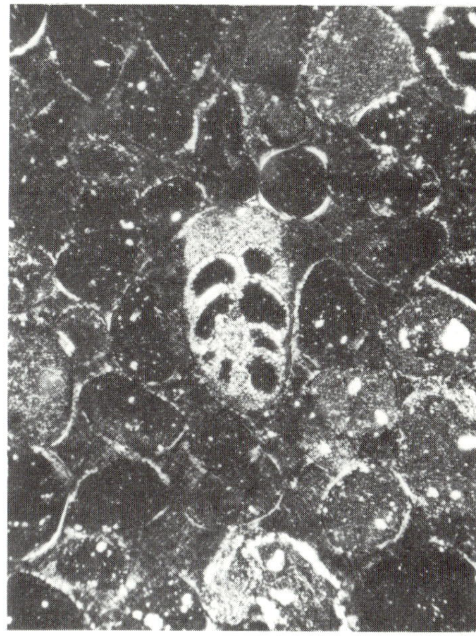

16b CP 0.1 mm

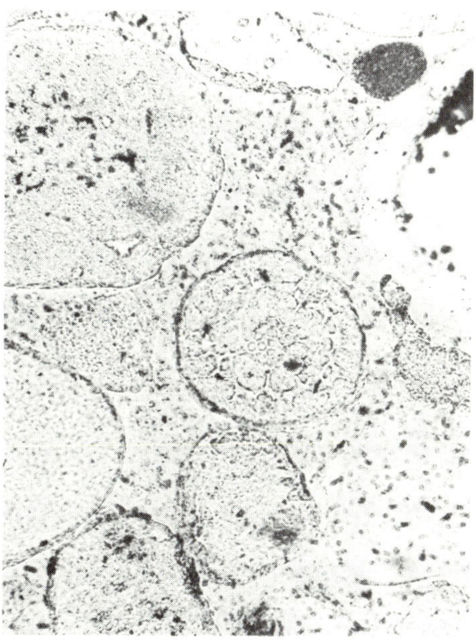

17 OL 0.1 mm

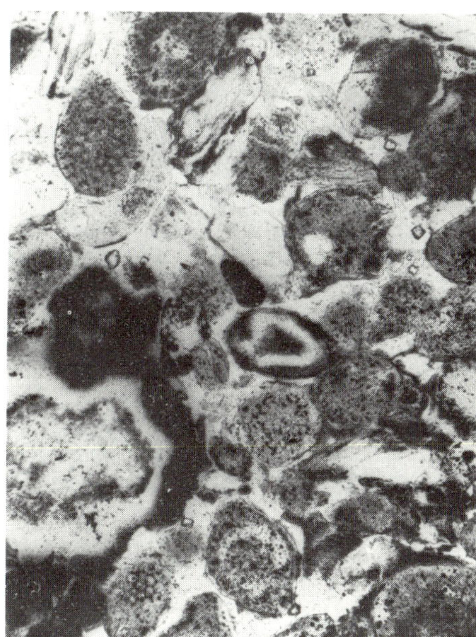

18 OL 0.1 mm

Plate 4 *Continued*

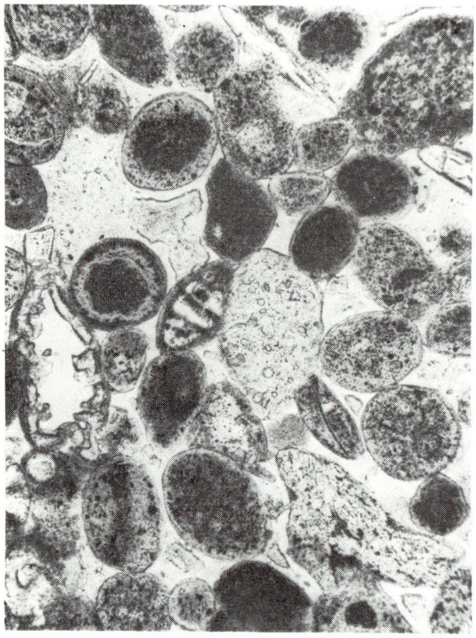

19 OL **0.1 mm**

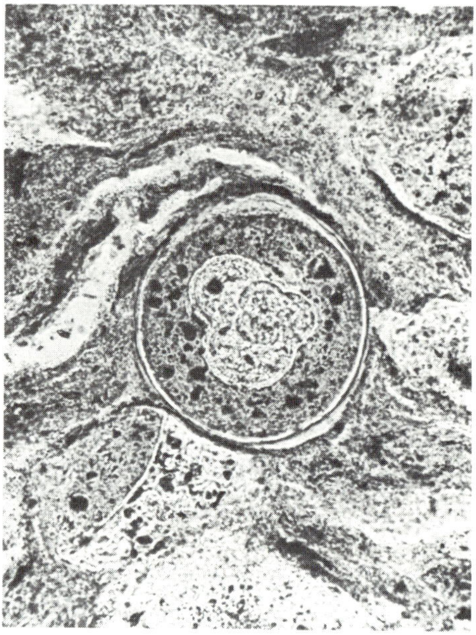

20 OL **0.1 mm**

OL = Ordinary light CP = Crossed polarized light

16a and **16b**: Pelphospharenite from the Bou Craa deposit. Foraminifera are converted by epigenesis into microcrystalline apatite, while pellets are cryptocrystalline. **17**: Same deposit. Pellet formed around a Radiolaria. **18**: Radiolaria pellets from the Middle Eocene in Benin. Top left: a coprolith. **19**: Pelphospharenite from the Kef Schfair deposit, Lower Eocene, Tunisia. The central grain is rich in diatom remains. **20**: Pellet centred on a foraminifera which has undergone epigenetic alteration; Middle Eocene, Benin.

of these layers are superimposed on one another around a single pellet, giving it an oolithic texture.

Although present in various phospharenites, well-defined ooliths are uncommon. According to Cook (1976), for example, they only represent 3% of the entire phosphate fraction of the Phosphoria Formation of the Permian of the Rocky Mountains.

2.2.2. Particles of rudite size

Particles of rudite size show greater variety in shape than arenite particles. They may be subdivided into bioclasts, intraclasts, coproliths and nodules.

Plate 5

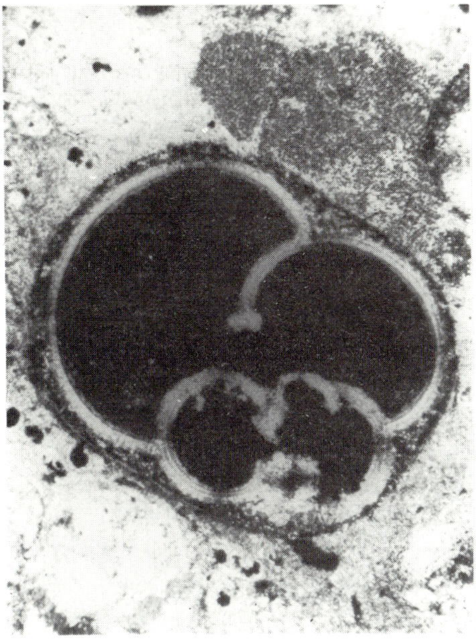

21 CP 0.1 mm

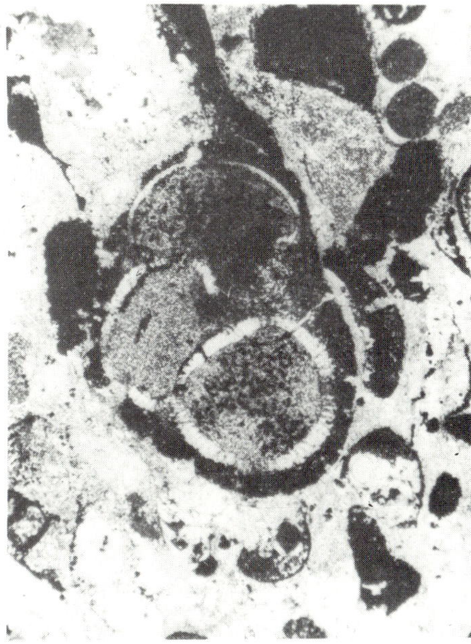

22 CP 0.1 mm

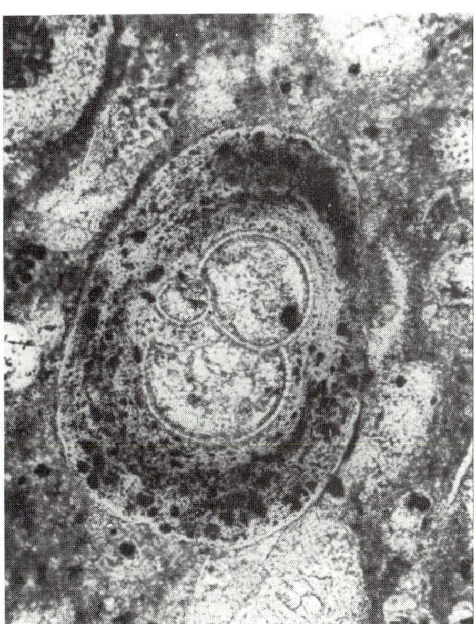

23 OL 0.1 mm

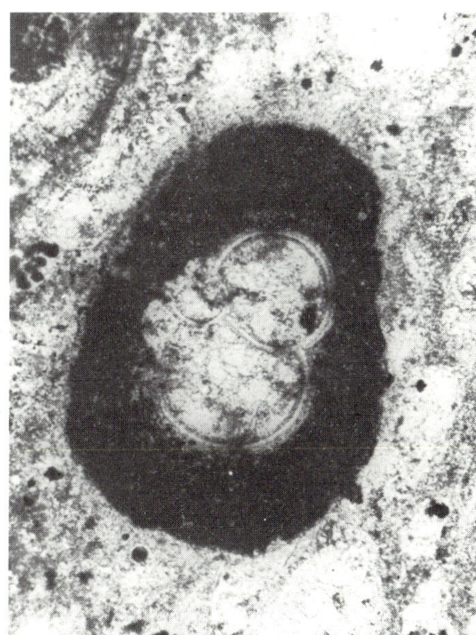

24 CP 0.1 mm

Plate 5 *Continued*

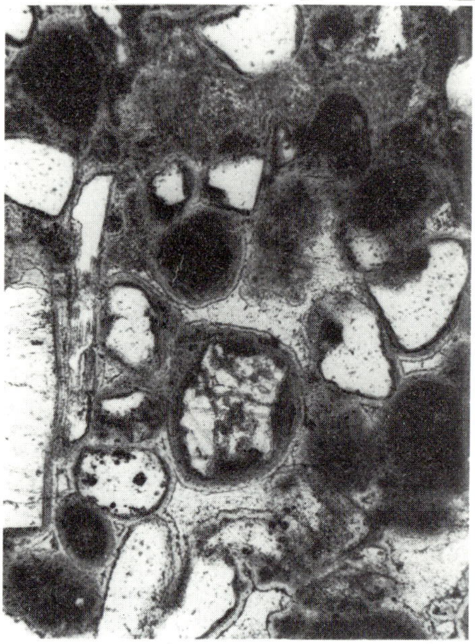

25 OL 0.1 mm

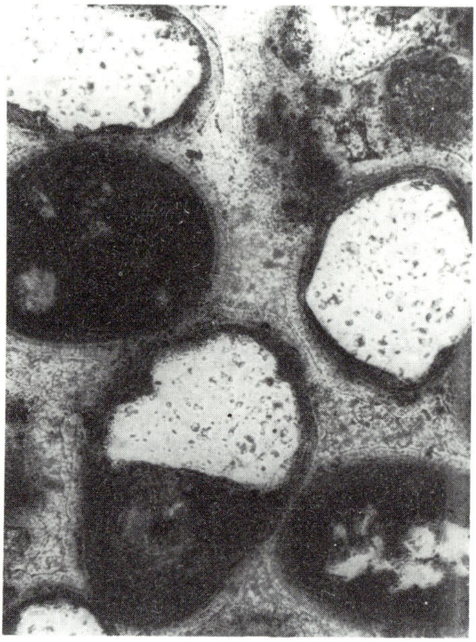

26 OL 0.1 mm

OL = Ordinary light CP = Crossed polarized light

21: Upper Eocene, Iran. Foraminifera with a calcite test, surrounded by microcrystalline apatite.

22: Same origin. Foraminifera with a calcite test.

23 and **24**: Foraminifera in calcite, at the centre of a pellet with an endogangue of organic matter.

25 and **26**: The Permian of the Rocky Mountains. Quartz and calcite grains surrounded by apatite, in a calcite matrix.

Bioclasts and intraclasts differ from arenite particles primarily by their size. Coproliths are of faecal origin by definition. This origin, however, is very difficult to confirm, so that opinions differ concerning which particles to include in this group. Certain authors (Riggs, 1979a; Champetier and Joussemet, 1979b; Porter and Robbins, 1981; Lamboy, 1982a) tend to generalize the coprolith concept, extending it to most pellets. Cayeux (1950), on the other hand, always uses this term with great care, confining it to cryptocrystalline elements of a general cylindrical shape, measuring from a few millimetres to 1 cm in diameter, and to certain elements of more varied shape but with a characteristic fluid internal texture.

Plate 6

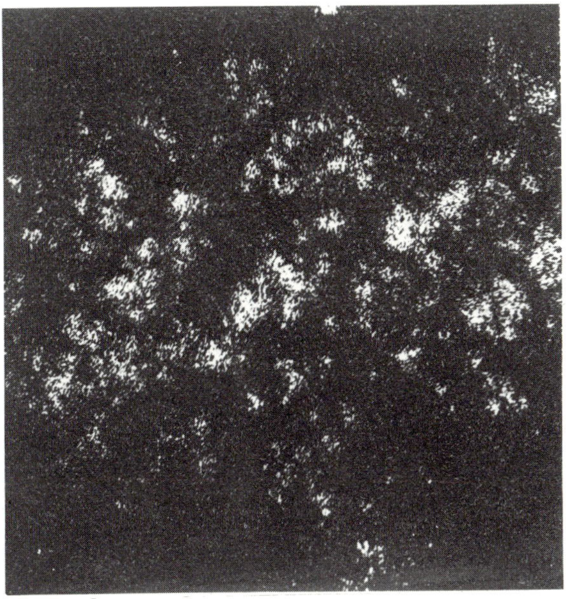

27 Phosphorus (× 800)

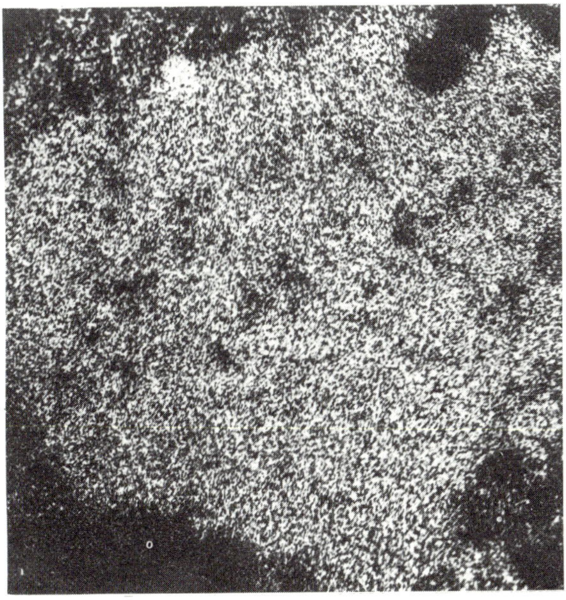

28 Potassium (× 800)

Plate 6 *Continued*

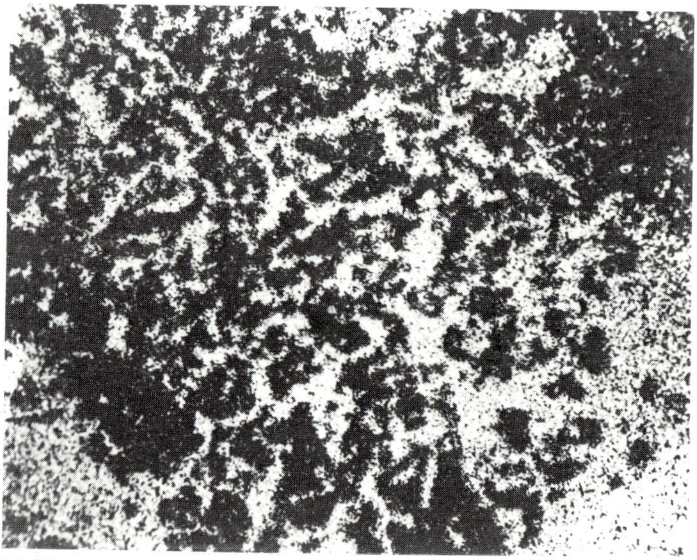

29 Phosphorus (× 400)

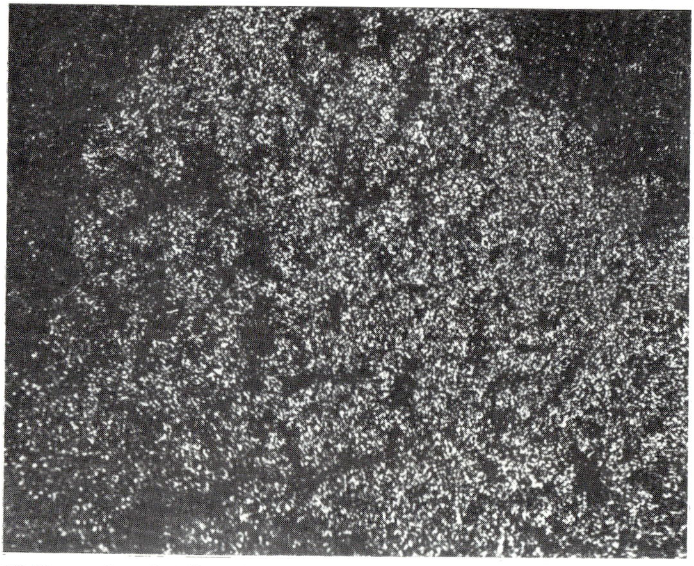

30 Potassium (400)

Study of two green phosphate grains from the Albian of the Paris Basin, using the Castaing electron microprobe (from Fauconnier, 1979).

27 and **29**: Phosphorus distribution within the apatite.
28 and **30**: Potassium distribution within the apatite.

	Probe analysis (area ~ 1μm²)			
	Upper Grain	Lower Grain		
SiO_2	33.46	35.85	42.43	53.50
Al_2O_3	4.66	4.51	5.08	6.90
Fe_2O_3	11.07	13.35	16.80	21.80
CaO	21.71	20.30	22.74	3.60
MgO	1.73	1.42	2.43	3.36
K_2O	4.74	4.32	5.22	6.47
P_2O_5	13.26	15.03	8.75	1.93

Present ideas discussed below on the environment in which the phosphate particles are formed, and the remarkable uniformity of shape of phosphate pellets throughout deposits ranging from the Infracambrian to the Pliocene period, necessitates equally careful use of the term. In petrographic work, the term coprolith should be avoided or used very sparingly. 'Phospharudite with cylindrical (or ovoid) granules (or gravel)' is preferable to 'conglomerate of phosphate coproliths', if there is any unresolved doubt as to the faecal origin of the phosphate particles.

The shape and size of the nodules are subject to considerable variation. The phosphate nodules of the Dinantian in the Pyrenees, for example, are black, flattened spherical or ovoid pebbles, or even cobbles, of microcrystalline apatite, with an endogangue rich in radiolarian tests and organic matter, often arranged in concentric bands (Plate 8, Photograph 40; Plate 9, Photograph 42).

The nodules of the Albian sequence in the Paris Basin are internal moulds of ammonites, pelecypods or gastropods, which have generally been reworked, or fragments of glauconitic sandstone with phosphate cement, which have also been reworked and sometimes occur as conglomerates. The nodules shown in Photograph 43 combine these facies. The pebbles found in Florida, according to Riggs (1979b), are intraclasts which are more or less rounded and polished by transport and which originate in a pre-existing deposit of microphatite, perforated by burrowing organisms.

3. THE EXOGANGUE

Certain deposits, such as those at Taïba (Senegal) and Bou Craa (Morocco), contain phospharenites with practically no exogangue; these unconsolidated very rich deposits, generally only extend for short distances vertically or hori-

Plate 7

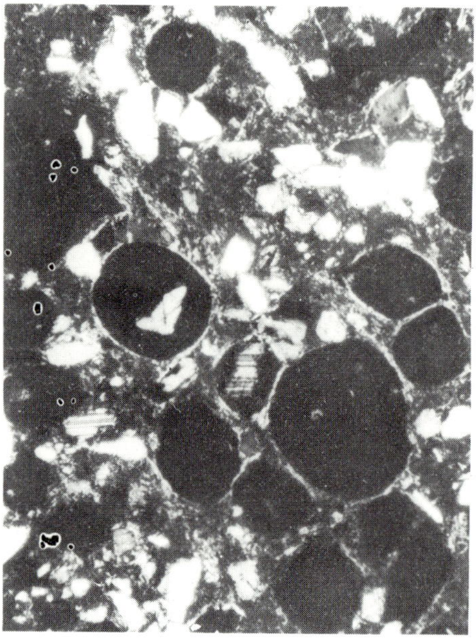

31 CP 0.1 mm

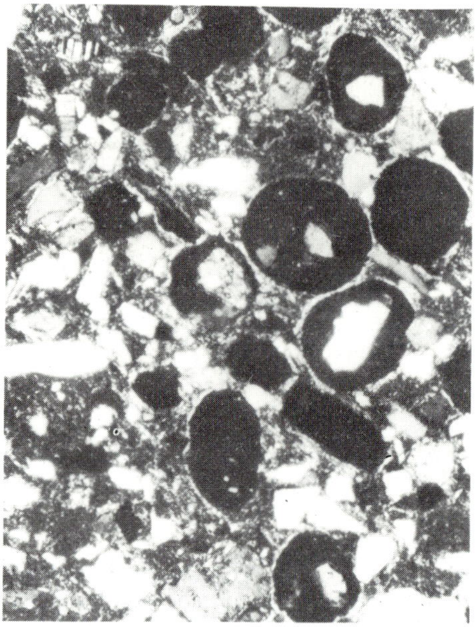

32 CP 0.1 mm

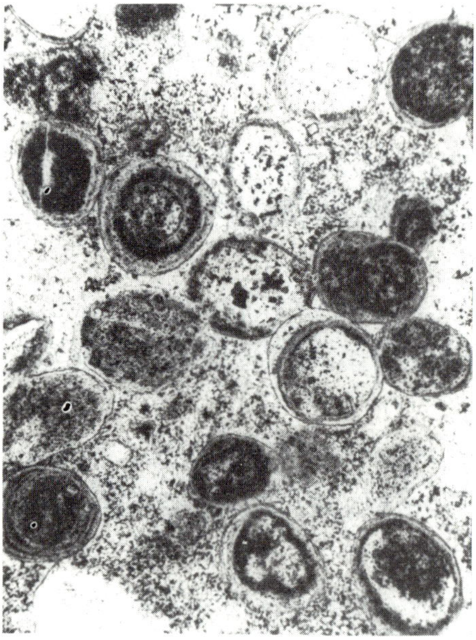

33 OL 0.1 mm

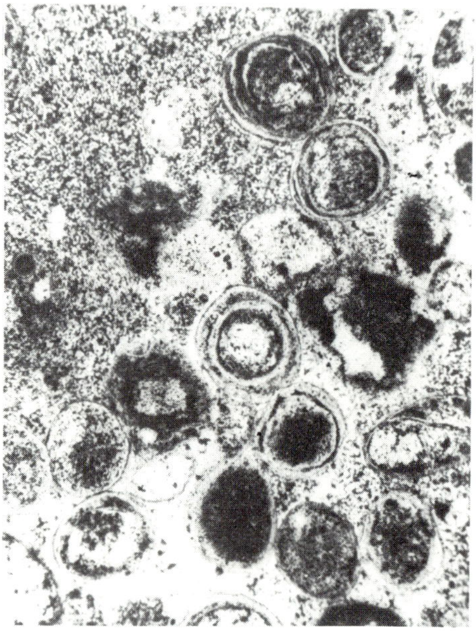

34 OL 0.1 mm

Plate 7 *Continued*

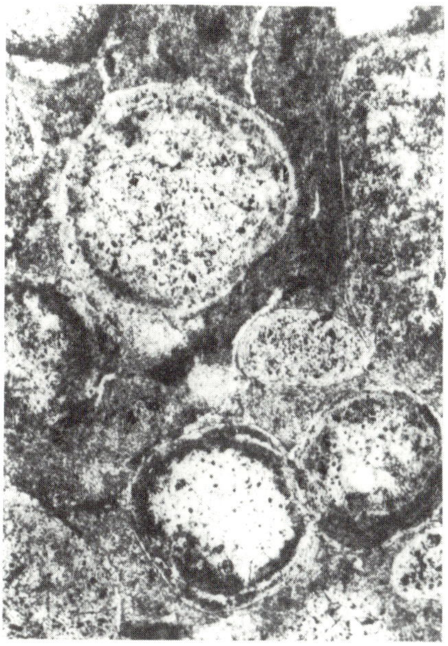

35 OL 0.1 mm

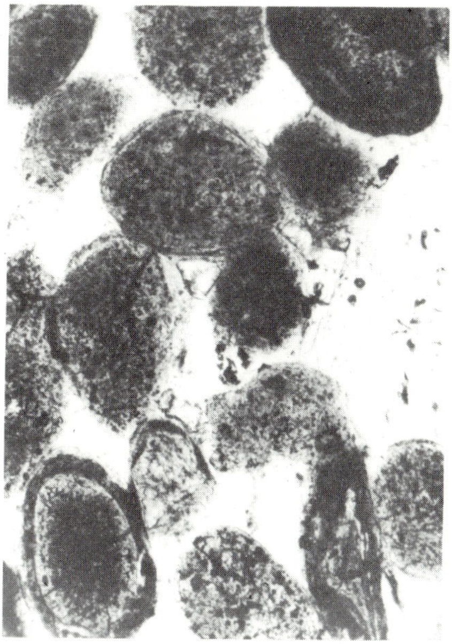

36 OL 0.1 mm

OL = Ordinary light CP = Crossed polarized light

31 and **32**: Californian Miocene. Pellets with a quartz and feldspar endogangue in quartzarenosiltite.

33 to **36**: Lower Eocene of the Gafsa Basin, Tunisia. Example of the distribution of organic matter in phosphate pellets.

zontally. The same may apply to beds of indurated phospharenite in which a subordinate fraction of microsphatite acts as a cement. In most cases, however, the exogangue is present and often abundant.

Clastic quartz is frequent in the exogangue of phosphatites presently mined. It is comparatively scarce in the Eocene deposits of North Africa, for example, but fairly abundant in the Eocene deposit at Taïba, abundant in at least part of the Infracambrian phosphate deposit of Niger, and very abundant in the Mio-Pliocene ores of Florida and California, which in many cases are no longer phospharenites but phosphatic quartzarenites. In the deposits, the quartz grains are generally smaller than the phosphate particles.

A cherty exogangue is found in deposits of the Upper Cretaceous or Tertiary age, but it is rare, and detailed study of the lithological series often reveals a

Plate 8

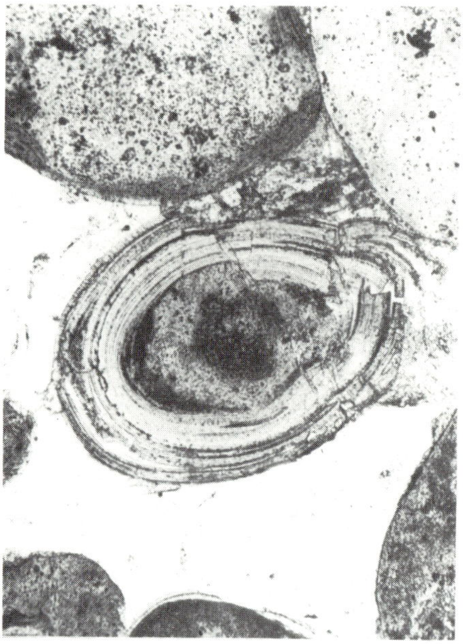

37 OL 0.1 mm

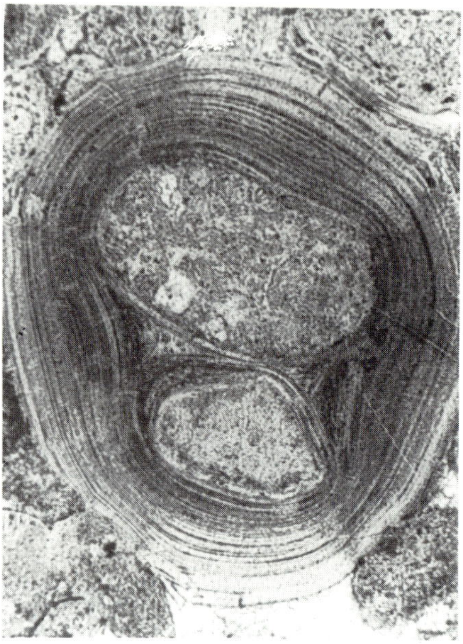

38a OL 0.1 mm

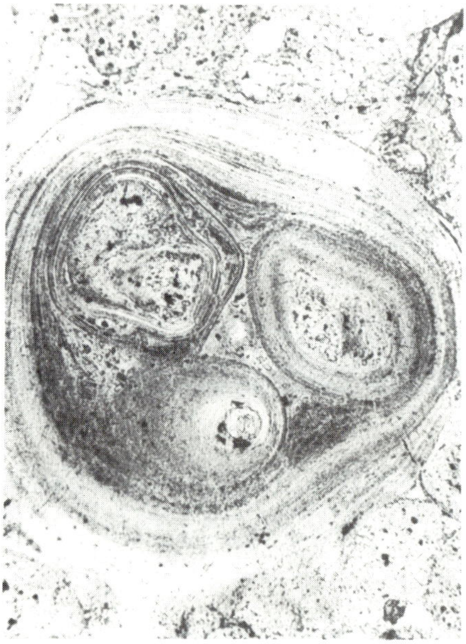

38b OL 0.1 mm

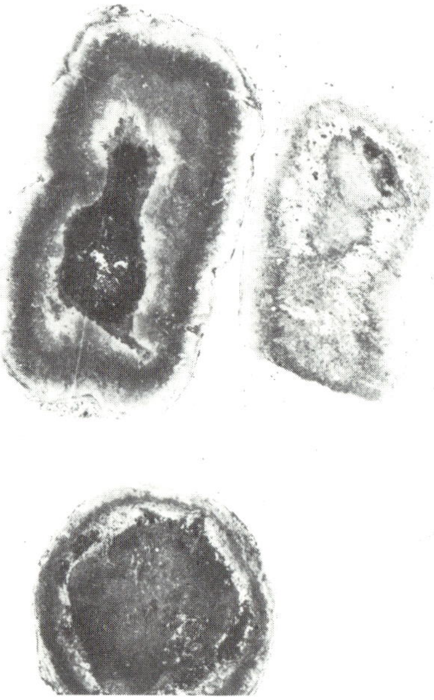

39 OL 1 mm

Plate 8 *Continued*

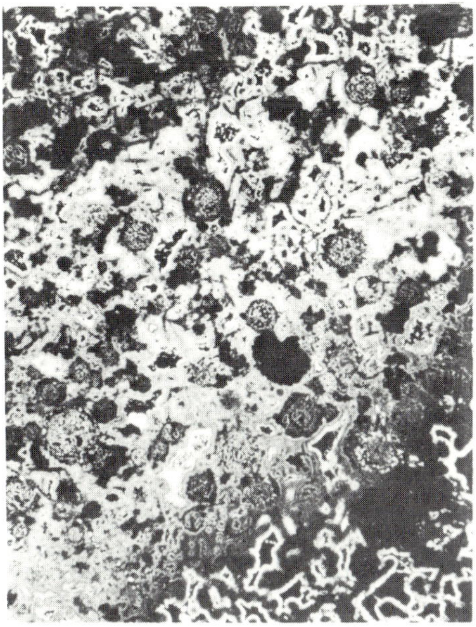

40 OL 1 mm

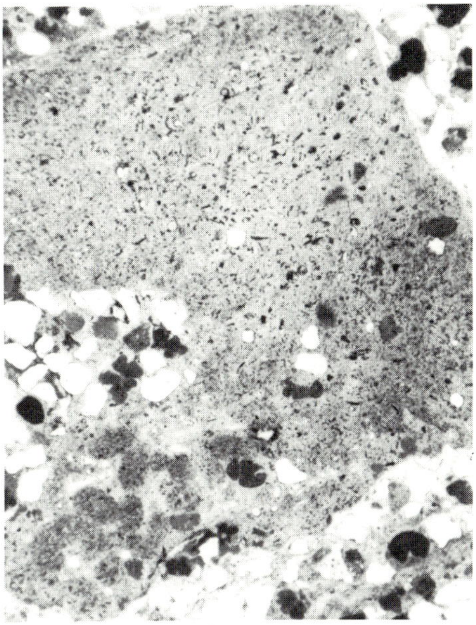

41 OL 1 mm

OL = Ordinary light

37, 38a and **38b** Infracambrian, Niger. Ooliths and compound pellets.

39: Upper Cretaceous coproliths. Angola (in Y. Champetier's collection).

40: Pyrenean Dinantian phosphate nodule, rich in radiolaria and organic matter (cf. Plate 9, Photograph 42)

41: Albian nodule of the north-eastern part of the Paris basin (cf. Plate 9, Photograph 43): phosphate concentrations containing sponge spicules, in a glauconitic sandstone with a cryptocrystalline apatite cement.

certain incompatibility between siliceous and phosphatic sedimentations. On the other hand, part of the gangue of the Precambrian deposits at Rajasthan, in India, and of the Infracambrian deposits of Niger, Burkina Faso and Benin have undergone more widespread, probably secondary, silicification. Phosphate nodules of the Dinantian of south-west France are closely associated with cherts.

Clay is frequently present in the exogangue of phosphatites, sometimes almost alone (e.g. in the Togo deposit) but often associated with a carbonate

Plate 9

42: Nodules containing 25 – 30% P_2O_5, from the Pyrenean Dinantian. The sectioned nodule to the left shows its concentric structure, highlighted by the variation in the concentration of organic matter. The one-Franc coin gives an indication of scale.

43: Nodules from the Albian of the north-eastern part of the Paris Basin, with variable P_2O_5 contents (5 – 25%) and a variety of shapes, many of them irregular. Some are internal moulds of pelecypods, gastropods or ammonites.

Plate 9 *Continued*

44: Pebbles from the Florida Mio–Pliocene phosphate formation. Riggs (1979a) believes they are intraclasts. They often contain quartz and their P_2O_5 content (30 – 32%) is generally lower than that of pellets.

fraction more or less rich in calcite and dolomite, as in the North African or Middle Eastern deposits. The most abundant clay minerals of the Upper Cretaceous or Tertiary deposits belong to the group known as smectites (montmorillonites). Kaolinite (Togo) or attapulgite and even sepiolite (Morocco, Tunisia) may also be present in varying amounts. The same applies to glauconite, but the best deposits, as already stated, are comparatively non-glauconitic. In the older phosphate deposits, the smectites are rare and they apparently never contain attapulgite or sepiolite. The main clay mineral appears to be illite.

A carbonate exogangue is very common in phosphate deposits and such occurrences will become more predominant as the low carbonate phosphate reserves, worked in preference because they are more profitable, are exhausted. In this exogangue it is generally calcite that predominates, but dolomite is often associated with it and is sometimes the most abundant mineral (e.g. part of the Precambrian deposit at Rajasthan, or phosphate deposits at the top of the Upper Cretaceous in the Masidagi region of Turkey).

The fractions referred to are most frequent in the exogangues of phosphatites, which may possibly also contain a greater or smaller quantity of iron: in sulphide, carbonate or oxide form, gypsum (e.g. in the Gafsa basin in

Tunisia) and sometimes barytes, pyroclasts, feldspars and various minerals of volcanic origin (e.g. in the Miocene in Baja, California). If there has been no excessive secondary oxidation, the exogangue is often rich in organic matter, particularly if it is argillaceous.

Chapter 4

Chemistry of phosphatites

1. MAJOR ELEMENTS

Table 9 summarizes the results of chemical analysis of samples of various ages, irrespective of whether they are representative of the phosphate depostis from which they are derived. These results were obtained at the B.R.G.M. or have been extracted from the literature. The following conclusions may be drawn from Table 9:

1. No relation can be established between the age of a sample and its chemical composition. This suggests that the conditions in which phosphatites are formed have remained reasonably stable with time.

2. The concentrations of SiO_2 are essentially due to the presence of silica or silicates, i.e. mainly of clays, including glauconites, and in some cases clinoptilolite. Feldspars are generally rare, except in the case of volcanic or terrigenous contamination, which may also lead to the presence of other silicates.

3. In fluorapatite $C_{10}(PO_4)_6F_2$, the CaO/P_2O_5 ratio is 1.31; it is higher in carbonate apatites, i.e. in most phosphatites. A lower ration than 1.3 indicates the presence of different phosphate minerals, generally aluminium or aluminocalcic. For example, the analysis of a sample taken from the weathered zone rich in augelite, crandallite and similar minerals, in the Taiba deposit in Senegal, gave the following results: P_2O_5: 28.7%, CaO: 9.8%, Al_2O_3: 31.5%, Fe_2O_3: 2.0%, CO_2: 0.5%, SiO_2: 6.8%, loss on ignition: 15.7%, MgO: 0.80%, F: 2.40%.

   The CaO/P_2O_5 ratio in this case is 0.34. It varies from 0.17 to 0.72 among the samples taken from this weathered zone (Slansky *et al.*, 1965, p.317) and reaches values above 1.3 in the underlying phosphatites.

4. If the CaO/P_2O_5 ratio is greater than 1.31, the increase may be due either to the substitution of PO_4 by CO_3 in the apatite, or to the presence of calcite or dolomite in the phosphatite.

Table 9. *Chemical analysis of phosphatite samples of various ages.*

Origin	Age	P_2O_5	SiO_2	Al_2O_3	Fe_2O_3	CaO	MgO
Peru–Chile (offshore)[a]	Holocene	22.61	22.13	5.15	2.85	33.93	1.07
Florida ('pebble')[b]	Pliocene	32.07	9.31	1.29	1.57	46.98	0.19
Venezuela (Riecito)	Miocene	34.28	7.05	1.00	0.69	48.05	0.23
Christmas Island[c]	Miocene	38.50		0.79	0.41	52.10	0.10
		37.40		3.10	1.17	48.60	0.15
Senegal (Taïba)	Mid. Eocene	33.30	7.30	3.20	3.60	45.10	0.60
Benin	Mid. Eocene	28.15	13.15	5.40	0.17	40.94	0.52
Togo (concentrate)	Mid. Eocene	36.85	2.99	1.00	1.30	51.69	0.03
Morocco (Khouribga)	Low. Eocene	34.26	0.03	0.37	0.26	52.78	0.48
(Bou Craa)		34.70	8.07	0.71	0.37	50.45	0.19
Tunisia (Metlaoui)	Low. Eocene	26.09	8.90	1.53	0.60	42.85	0.50
		24.70	7.80	0.52	0.71	44.16	3.04
Egypt (Nile Valley)	Up. Cretaceous	26.00	9.25	0.53	1.77	45.14	0.77
(Quseir)		25.20	12.50	0.84	1.86	40.66	1.75
Israel (Oron)[d]	Up. Cretaceous	25.20	2.00	0.50	0.30	52.50	0.20
Syria	Up. Cretaceous	24.50	6.90	0.08	0.25	44.16	0.27
Colombia	Up. Cretaceous	28.04	17.40	3.10	1.72	40.08	
Thailand (Maeta)	Permian	38.84	0.01	0.51	0.97	51.10	0.10
India (Mussoorie)	Permian	22.50	7.05	0.66	2.56	40.55	6.00
U.S.A. (Rocky Mountains)[e]	Permian	30.50	11.90	1.70	1.10	44.00	0.30
Australia (Duchess)[f]	Cambrian	37.20	2.60	0.85	0.94	51.70	0.15
(Lady Annie)[g]		35.00	10.30	1.59	0.14	48.20	0.15
China[h]	Up. Cambrian	23.41	19.74	3.12	1.90	41.98	0.21
Benin (Mekrou)	Inf. Cambrian	28.25	24.65	1.28	2.14	38.55	0.08
India (Udaipur)[i]	Precambrian	26.12	20.93	3.02		40.28	
		35.46	10.56	1.18		47.75	0.28

[a] Burnett, 1974, p. 105.
[b] Altschuler *et al.*, 1964, p. 31.
[c] Barrie, 1967, Table 1.
[d] British Sulphur Corporation Ltd, 1980, p. 163.
[e] Gulbrandsen, 1966, Table 1, mean of 63 tests.
[f] Russel and Trueman, 1971, p. 1210.

Na_2O	K_2O	CO_2	SO_3	F^-	Cl^-	H_2O^-	H_2O	C org.	$\frac{CaO}{P_2O_5}$	$\frac{F}{P_2O_5}$	$\frac{CO_2}{P_2O_5}$	$\frac{SO_3}{P_2O_5}$
0.85	1.30			2.22					1.50	0.098		
0.21	0.13	3.07	0.59	3.68	0.013	1.88		0.053	1.46	0.115	0.09	0.018
0.70	0.08	3.00	0.70	2.23	0.08	0.42	1.81	0.09	1.40	0.065	0.08	0.02
0.20	0.03	1.20		1.14		1.43	1.97		1.35	0.029	0.03	
0.22	0.06	2.00		2.17		1.73	3.19		1.29	0.058	0.05	
		1.40		3.75		0.60			1.35	0.112	0.04	
0.30	0.21	1.79		2.65	0.48	1.77	3.45	0.07	1.45	0.094	0.06	
0.27	0.05			3.75	0.12	1.44			1.40	0.101		
0.84	0.09	3.59	1.59	3.05	0.03		2.34		1.54	0.089	0.10	0.04
0.07	0.10	1.98	0.66	3.56		0.74			1.45	0.102	0.05	0.019
1.45	0.38	4.62	3.90	2.98	0.09	3.16	3.00	0.93	1.64	0.114	0.17	0.149
1.23	0.19	10.95		2.92	0.24	0.42	1.46	0.34	1.78	0.118	0.44	
0.78	0.12	8.92		2.57		0.66	1.73	1.32	1.73	0.098	0.34	
0.68	0.10	5.58		2.57	1.06	2.11	2.19	1.20	1.61	0.101	0.22	
0.80	0.03	13.00	1.80	2.90	0.40				2.08	0.115	0.51	0.07
0.52	0.03	2.92	11.20	2.65	0.26	4.65	1.90	0.26	1.80	0.108	0.12	0.45
		2.77		1.98					1.42			
0.10	0.03	2.40	0.13	0.78	0.06	0.49	2.26	0.06	1.31	0.020	0.06	0.003
0.18	0.24	15.10	1.35	2.15	0.06	0.24	0.75	0.76	1.80	0.095	0.67	0.06
0.60	0.50	2.20	1.80	3.10		0.60	1.60	2.10	1.44	0.101	0.07	0.059
0.28	0.09	1.63	0.56	3.20		0.53	1.42		1.38	0.086	0.04	0.015
0.04	0.09	1.20	0.07	3.16		0.53	0.70	0.60	1.37	0.909	0.03	0.002
0.54		5.52		2.72	0.67	0.33	0.24	0.62	1.79	0.116	0.23	
0.18	0.10	1.12		2.60	0.05	0.21	1.29	0.07	1.36	0.092	0.03	
		4.03		2.01		0.46			1.54	0.076	0.15	
		0.94	0.21	1.60		1.66			1.34	0.045	0.02	0.05

[g] Rogers and Reevers, 1976, p. 259.
[h] Altschuler, 1973, p. 43.
[i] Banerjee, 1971, p. 2326.

Both effects may act together, up to a certain level at least, as has been demonstrated by Gulbrandsen (1970) in the Phosphoria Formation of the Rocky Mountains, where the CO_2 content of the apatite is highest in areas in which carbonates are abundant and lowest where they are absent. The limit of the CaO/P_2O_5 ratio seems to be about 1.6; chemical analysis of commercial concentrates carried out by Lehr *et al.* (1967) after extraction of the free carbonate shows that in fact the ratio in the sample with the highest CO_3/PO_4 molar ratio (Gafsa concentrate) is 1.57. It is therefore only above this value that the CaO/P_2O_5 ratio is definitely indicative of the presence of free carbonates or possibly of other minerals (gypsum or even certain clays).

5. Similar reasoning also applies to the CO_2/P_2O_5 ratio, in which increases of up to about 0.18 may be due both to the composition of the apatite itself and to the presence of free carbonate.

6. Fluorine may also appear in phosphatites in a form (usually fluorspar) independent of apatite, but this is rare. Generally, fluorine is bound to the apatite, and the F/P_2O_5 ratio, which is therefore a characteristic of the mineral, varies between 0.08 and 0.12, occasionally reaching 0.13 or even 0.14. This ratio is far lower in deposits derived from guano, as on Christmas Island or on Curaçao, for example, or in certain deposits where the phosphate has been affected by secondary solution, as at Mae Ta in Thailand, or the Miocene deposit at Riecito and Venezuela. The low F/P_2O_5 ratio of the Precambrian samples from Udaipur (India) is probably also due to their comparable history.

7. The sedimentary apatites are usually very poor in chlorine. Where relatively high quantities are found (e.g. at Quseir), they are generally the result of secondary concentration outside the apatite lattice and can be reduced by fresh-water washing.

8. Sulphur associated with phosphatites may appear in sulphides, sulphates or organic matter. Belayouni (1984) has shown that organic sulphur is generally present in humic acids in the phosphate grains and in kerogene in the exogangue. In phosphate-rich strata, the proportion of organic sulphur by weight in the total rock does not appear to exceed 0.2–0.3% (i.e. 0.5-0.7% SO_3); it may exceed 1% (2.5% SO_3) in strata poor in phosphate but with a high organic matter content.

   The sulphides are generally present at pyrites and marcasite, both in the exogangue and in the endogangue; their presence is indicated by the corresponding values of iron. The most common sulphate is gypsum; there may be small amounts of baryte or celesite. According to Duchatelle (1969), sulphates are only present in the exogangue — SO_4 ions inside the phosphate pellets are associated with the apatite lattice. According to Sassi (1974), the weight ratio of SO_3 to PO_4 may reach the order of 0.12 in francolite of the Gafsa basin, but it is often lower, e.g. 0.03-0.06, in Moroccan samples.

9. Aluminium and iron, usually expressed in terms of oxides in analysis, play a significant role in the use of phosphate ores. These elements are principally associated with clays, sulphides, hydroxides or even iron and aluminium phosphates. The combined Al_2O_3 and Fe_2O_3 concentration is often called 'Feral' by French-speaking experts, while English-language literature generally uses the expression R_2O_3.

2. TRACE ELEMENTS

Information on phosphatite trace elements is still fairly limited. The most complete data available were published in 1966 by Gulbrandsen, with reference to the Permian Phosphoria Formation of the Rocky Mountains, from an analysis of 62 samples. The average contents of the major elements are given in Table 9; those of the trace elements are given in Table 10, expressed in ppm.

These contents may, however, vary very widely. For example, U content varies from 30 to 130 ppm, Ag from 0.1 to 10, As from less than 10 to 200, Ba may reach 300, Be may reach 30 and Cd 300, Co only reaches 30 ppm in one case and generally remains below 10; Cr varies from 300 to 3000 ppm, Cu from 10 to 300 ppm; the maximum for Ga is 10 ppm; La varies from less than 30 to

Table 10. *Mean trace element contents (ppm) of Permian phosphatite samples, Rocky Mountains* (from Gulbrandsen, 1966).

U	90	Mo	30
Ag	3	Nd	300
As	40	Ni	100
B	<50	Pb	<10
Ba	100	Sb	7
Be	<0.5	Sc	10
Cd	<50	Se	13
Co	<10	Sr	1000
Cr	1000	V	300
Cu	100	Y	1000
Ga	<10	Yb	10
La	300	Zn	300
Mn	30	Zr	30

300, or even 1000 ppm in one case; Mn may reach 300 ppm, Mo 100 to 300; Na, often found in concentrations of 300 ppm, may be far less plentiful; Ni generally varies from 30 to 300, Pb may reach 30 ppm at the most, and Sb 15; Sc varies from below 5 to 30 ppm, Se from 1 to 70; Sr, often in concentrations of 1000 ppm, may fall as low as 300 ppm; V varies from 100 to 1000 ppm; Y, often in concentrations of 1000 ppm, falls to around 300 ppm; Yb varies from 3 to 30 ppm; Zn often reaches 300 ppm, but may be far lower; Zr varies from 10 to 100 ppm.

The contents of seven trace elements established in 1961 by the B.R.G.M. in samples of ore from Taïba in Senegal were comparable with Cr, U, Cu, Ni and Co; higher figures were obtained for Pb (260 – 320 ppm) and Zn (600 – 1000 ppm). Note, however, that care is needed in comparing contents of trace elements found by different methods or in different laboratories.

According to Kholodov and Tooms *et al.* (in Altschuler, 1973, p. 42) the concentrations of B, Ba, Be, Co, Li, Pb, Ni, V, Zn and all the rare earths are generally higher in phosphate rock than in the Earth's crust; and the concentrations of Ag, As, Cd, Cr, I, La, Mo, Pb, Sb, Se, Sn, Sr, U and Y may even be twice as high.

According to Prévôt and Lucas (1979), the most common trace elements of the phosphatites are Sr, Ba, V, Ni, Cr, Zn, Cu, Mo, Ag, Se, U and the rare earths. For seven groups of phosphate specimens from the Moroccan deposits they obtained the results shown in Table 11.

According to Krauskopf (1955) or Gulbrandsen (1966), the abundance of the trace elements in phosphatites is due to their association with organic matter, or to their integration into the apatite lattice. Concentrations of Ag,

Table 11. *Trace elements in Moroccan phosphatites* (from Prévôt and Lucas, 1979).

Element	Contents (ppm)		
	minimum	maximum	mean
Sr	617	1288	974
Ba	64	155	104
V	42	191	98
Ni	21	64	43
Cr	149	298	220
Zn	144	279	213
Cu	22	56	39

As, Mo, Ni, V, Zn, Cr, Cu, Sb, Se and Cd are principally due to organic matter, while those of Sr, U, Th and the rare earths are due to the apatite itself. Vanadium, however, may also be associated with the apatite, while uranium may be present in organic matter. Prévôt and Lucas (1979) add that there is a possible association with clays, e.g. in the case of Ti.

3. CONCENTRATION OF URANIUM IN PHOSPHATITES

The uranium content of phosphatites is very variable.

3.1. Most frequent concentrations

In most phosphate deposits the uranium content of the ore is often below 100 ppm and only rarely exceeds 200 ppm, as indicated by the following examples:

1. In the Permian Phosphoria Formation (Gulbrandsen, 1966) of the Rocky Mountains (U.S.A.), the average uranium content for 62 samples is 90 ppm, with a maximum of 210 ppm, for an average of 30.5% P_2O_5, i.e. an average of 3 ppm of uranium for every 1% of P_2O_5.

2. In the Egyptian deposits of Upper Cretaceous age (Zaghloul and Mabrouk, 1964), uranium contents of 30 – 100 and 98 – 190 ppm have been found for the Red Sea deposits; for those of the Nile Valley, various authors mention 82, 97, 68, 150 and 130 ppm.

3. In Jordan (Bashir, 1977), 26 samples distributed among the three deposits of Ruseifa, Al-Hasa and Shadya (Upper Cretaceous) have been found to contain between 14 and 156 ppm of uranium for between 5 and 34.3% P_2O_5. The U(ppm)/P_2O_5(%) ratio varies from 1.4 to 12, with an average close to 3, except for a group of samples for which it is as high as 9.

4. In Syria, deposits of the same age as in Jordan have similar uranium contents; the highest contents are apparently associated with the highest P_2O_5 concentrations (e.g. 130 ppm for rock with 35% P_2O_5); the converse does not necessarily hold, however, and a sample with 35% P_2O_5 may contain only 60 ppm of uranium. These two examples show that the U(ppm)/P_2O_5(%) ratio may vary from 1.7 to 3.7.

5. In Israel, Nathan and Shiloni (1976) show from a study of 254 samples that the uranium contents in phosphates are subject to regional variations. The Zefa deposits, for example, contain between 115 and 126 ppm of uranium for samples with P_2O_5 concentrations ranging from 20 to 36%. The U(ppm)/

P_2O_5(%) ratio varies from 4.15 to 10.97, with an average of 6. For the Saraf deposit, uranium contents vary from 67 to 132 ppm and the P_2O_5 content from 21.2 to 29.4%. The U(ppm)/P_2O_5(%) ratio lies between 2.97 and 4.61, with an average of 3.5.

6. In Tunisia, uranium contents do not exceed 60 ppm in the Gafsa Basin (Sassi, 1974), but reach about 100 ppm in deposits further north. Analysis of the samples obtained by drilling in Kef Schfair (Fauconnier and Slansky, 1980) show a U(ppm)/P_2O_5(%) ratio of between 0.6 and 1.3 for P_2O_5 concentrations exceeding 20%.

7. In Turkey, the ore of the Mazidagi region (Arda *et al.*, 1976) shows a uranium content that only reaches or exceeds 100 ppm in six of the 228 samples analysed. The U(ppm)/P_2O_5(%) ratio averages about 2.5.

8. In Morocco, according to Ziegler (written communication, 1978), the average uranium content in the various strata depends on their age. Estimates suggest that uranium contents reach 130 ppm in Ypresian Bed I at Khouribga, 160 ppm in Thanetian Bed II and 190 – 250 ppm in Maastrichtian Beds III and IV. This increase in the uranium content is independent of the P_2O_5 content, which is at its maximum in Ypresian Bed I and at its minimum in the Maastrichtian Bed. Chambionnat (1963) confirms that the highest levels are in the deepest strata, but draws attention to considerable lateral variations in the more recent strata.

9. In Florida (Miocene and Pliocene), the uranium contents regarded as typical (Altschuler *et al.*, 1958) of the different facies are 61 ppm (Pellets, Hawthorn formation), 90 ppm (Pellets, Bone Valley formation), 160 ppm (Pebbles, Bone Valley formation).

Table 12. *Uranium contents of Recent phosphate nodules* (from Baturin and Kochenov, 1974).

	Mean values			Number of samples
	P_2O_5%	U (ppm)	U (ppm)/P_2O_5 (%)	
Lenses of phosphatized diatomaceous ooze	8.32	5	0.5	3
Unconsolidated concretions	25.20	14	0.6	3
Compacted concretions	29.37	35	1.4	6
Dense concretions	33.09	53	1.6	5
Unconsolidated coproliths	25.85	28	1.0	2
Compacted coproliths	31.64	68	2.2	1
Dense coproliths	32.06	77	2.4	1

10. In South America, Cathcart (1978) found averages of 80 ppm in the ore of Upper Cretaceous age in Colombia, 60 ppm in the Miocene ore of Peru and 20 ppm in the Infracambrian ore of Brazil (Bambui).

11. Phosphorites found on the present ocean floor have uranium contents comparable with those of ores found on the land masses. In recent nodules, uranium content increases with the degree of lithification; the western continental margin of southern Africa gave the results shown in Table 12 (Baturin and Kochenov, 1974).

These values, which are close to those found in geologically older ores, indicate an increase in uranium content of the phosphate deposits during diagenesis.

3.2. Higher concentrations

Certain deposits contain far higher uranium concentrations than those considered so far. In the Eocence deposit of Bakouma (Central African Republic), the 15 samples studied by Gony (1971) contain from 1660 to 5600 ppm for a P_2O_5 concentration between 8.75 and 33.58% (average: 20.37%), as shown in Table 13. The U(ppm)/P_2O_5(%) ratio varies from 81.7 to 392.9, with an average of 155.7. The uranium content of the apatite is therefore between 10 and 100 times greater in this deposit than in most others. Note that the origin of the Bakouma phosphate deposit is still the subject of debate; Bigotte and Bonifas (1968) regard it as marine, while Miauton (1980) suggests a continental origin.

The Mussoorie deposit in India, of Palaeozoic age, also contains unusually high amounts of uranium (Saraswat *et al.* 1970), although lower than at Bakouma. For P_2O_5 contents varying from 7.19 to 23.62% (with a mean over 24 samples of 15.34%), uranium contents vary from 25 to 831 ppm (average: 408 ppm) and the U(ppm)/P_2O_5(%) ratio ranges from 2.2 to 50.5 (average 29.0).

Other high concentrations of uranium are reported (Gabshin *et al.*, 1974) in the Middle Palaeozoic deposits of Siberian basins with a high pyroclastic content: phosphatic lenses with cryptocrystalline carbonate fluorapatite, associated with fish remains, contain from 1 to 20% P_2O_5, with occasional maxima of 24%, and from 100 to 4000 ppm of uranium. Samples with over 10% P_2O_5 all contain more than 1000 ppm of uranium. High concentrations of uranium are also associated with the phosphatic strata in the Eocene lacustrine sediments in Wyoming and Utah (up to 3000 ppm), with certain Miocene phosphatic sandstones in Japan (up to 1200 ppm) and with the Visean phosphatic limestones of northern England (up to 700 ppm).

Table 13. *Distribution of the three forms of uranium in the Bakouma deposit (Central African Republic), according to Gony (1971), who differentiates between a 'transitional zone' to a depth of 25 m, and a 'lower reducing zone' to 35 m.*

	Weighted percentages												Contents (ppm)				
	No. of samples	CaO	P_2O_5	CO_2	F^-	SiO_2	Al_2O_3	Fe^3	Fe^2	Na_2O	K_2O	MgO	Total U	Total U^{6+}	Free U^{6+}	Bound U^{6+}	U^{4+}
Upper zone	11	10.50	8.75	0.57	0.61	58.02	6.70	2.91	0.21	0.27	3.13	1.24	1 660	1 050	350	700	610
	12	30.45	23.15	1.67	2.03	27.38	2.85	5.46	0.12	0.59	0.63	0.53	5 140	4 160	1 250	2 910	980
	13	15.22	11.95	0.78	0.85	48.52	6.65	4.19	0.11	0.44	2.17	0.91	3 500	2 640	1 550	1 090	860
	15	17.63	14.25	0.84	0.94	45.50	7.97	2.86	0.08	0.35	2.63	0.93	5 600	5 300	2 900	2 400	300
Transitional zone	2	36.75	25.98	2.20	2.09	18.63	2.73	3.60	0.21	0.75	0.43	0.48	3 140	2 200	300	1 900	940
	8	19.95	15.58	1.04	1.41	45.13	6.30	2.61	0.15	0.29	2.49	0.91	1 890	1 250	320	930	640
	9	17.85	13.18	0.91	1.12	49.52	5.07	2.15	0.17	0.39	2.43	1.04	1 680	1 250	260	990	430
	10	19.77	15.33	1.02	1.17	44.90	5.15	3.77	0.12	0.25	1.58	0.83	2 100	1 390	450	940	710
Lower zone	1	39.03	28.76	2.33	2.07	17.44	1.91	2.29	0.29	0.82	0.28	0.08	3 360	2 300	90	2 210	1 060
	3	33.60	24.38	2.07	2.22	26.51	2.83	1.72	0.24	0.67	0.54	traces	3 000	1 930	90	1 840	1 070
	4	28.87	20.55	1.55	1.41	36.22	4.20	2.72	0.18	0.70	0.93	0.41	2 520	1 760	130	1 630	760
	5	31.85	22.63	2.02	1.78	30.15	3.31	2.39	0.23	0.77	0.57	0.50	1 850	1 300	100	1 200	550
	6	22.05	16.25	1.18	1.33	45.56	4.00	1.83	0.15	0.44	1.07	0.65	1 720	920	60	860	800
	7	42.73	31.30	3.10	2.00	11.07	0.10	2.02	0.08	1.37	0.13	0.02	2 300	1 900	140	1 760	400
	14	44.24	33.58	3.43	2.04	5.92	0.20	1.62	0.08	1.20	1.14	0.15	3 680	2 570	130	2 440	1 110

3.3. Types of association between uranium and phosphatites

All modern studies agree in associating uranium directly with apatite in sedimentary phosphate deposits. Surface weathering of ore may, however, produce other uranium-bearing minerals (vanadates, autunite, etc.) which, except in the particular case of Bakouma, generally only appear as rare coloured coatings. Where lateritic weathering has destroyed the apatite, uranium appears to be primarily associated with the alumino-calcic phosphate minerals rather than with those that are exclusively aluminous.

Locally, the uranium content often increases with the corresponding P_2O_5 content, but it is only with very low amounts of uranium that the relationship seems to be really linear, as shown by Coppens *et al.* (1977) in connection with samples of Jordanian phosphate: up to 50 ppm of uranium, the U(ppm)/ P_2O_5(%) ratio remains between 2 and 3, but beyond 50 ppm it varies from 3.6 to 12 in an apparently random manner. These authors believe that some of the uranium might be associated with organic matter or with clays. Similar ideas have been advanced by Berzina *et al.* (1976) in connection with the Vyatka-Kama deposit (U.S.S.R.).

As has been demonstrated by Altschuler *et al.* (1958), uranium is associated with sedimentary apatites in two forms: on the one hand, U^{4+} partially replaces Ca^{2+} in the lattice, mainly because of their comparable ionic radii (0.97 and 0.99 Å); on the other hand, U^{6+}, which cannot be incorporated into the lattice, is adsorbed on the surface of the crystal. The tetravalent form of uranium appears to predominate at first, at the moment when the original crystallization of apatite occurs, whereas the hexavalent form emerges later, as a result of oxidation; it appears, however, that the two forms have coexisted from the beginning.

In Florida (Altschuler *et al.*, 1958), U^{4+} represents 40 – 91% of the total uranium in the Hawthorn and Bone Valley formations; it represents 55 – 74% of the total uranium in post-Miocene phosphate nodules of the Pacific. Its percentage is lower in older phosphates: 8 – 15% in the Senonian samples of Morocco, 18 – 24% in samples of the Phosphoria Formation, etc. At Bakouma, the average U^{4+} content is one-third of the total uranium; in this deposit, however, some of the U^{6+} is no longer associated with the apatite near the surface and enters into autunite, meta-autunite. Table 13, based on data supplied by Gony (1971), shows the distribution of these three forms of uranium in different samples.

Uranium in phosphates generally appears to be in radioactive equilibrium. This has allowed ocean floor phosphate deposits to be dated, confirming that phosphate sedimentation is still taking place today, which was previously doubted. However, Coppens *et al.* (1977) have shown that a collection of samples of Jordanian ore is in radioactive disequilibrium: many of the decay members of U^{238} are missing. The practical consequences of this disequilibrium, from the prospector's viewpoint, is that radioactive levels are not an indication of the uranium content in the region.

4. RARE EARTH CONCENTRATIONS IN PHOSPHATITES

The average rare earth contents (yttrium and lanthanides) vary from 25 ppm in limestone to over 250 ppm in shales (Handbook of Geochemistry, 1970), with a general average of 170 – 230 ppm (mainly 150 – 200 for the lanthanides alone). According to Baturin *et al.* (1972), the contents of recent phosphorites are lower; concretions with 27.7 and 25.6% P_2O_5 obtained on the Namibian and Peruvian continental margins, respectively, contain less than 50 ppm and about 100 ppm, respectively, of rare earths. The older phosphate particles from the present ocean floor show higher concentrations: 330 ppm (19.1% P_2O_5) for a sample taken at a depth of 500 m on the Agulhas Bank, 400 ppm (20.3% P_2O_5) for a sample taken at a depth of 300 m on the Blake Plateau, 660 ppm (21.3% P_2O_5) for a sample taken at a depth of 130 m off the Namibian coast and 980

Table 14. *Lanthanide, yttrium and scandium contents of phosphatites* (from Altschuler, 1973; El Kammar *et al.*, 1979).

Element	U.S.S.R.	Florida	Nile Valley
La	1 500	150	134
Ce	2 700	120	154
Pr	500	30	38
Nd	1 500	70	118
Sm	430	30	22
Eu	20	4	4.4
Grd	330	14	23
Tb	40	4	3.2
Dy	300	16	22.1
Ho	50	4	4.8
Er	220	21	15.7
Tm		2	1.6
Yb	100	8	10.5
Lu	N.D.	3	2.4
Total lanthanides	7 690	476	553.7
Y	1 750	110	210
Sc	70	3	N.D.
Total	9 510 ppm	589 ppm	763.7 ppm

ppm (30.9% P_2O_5) for a block obtained from a depth of 1950 m on a guyot in the Pacific.

Concentrations of rare earths vary considerably in phosphate deposits: 590 ppm in Florida (Altschuler *et al.*, 1967), 128 – 691 ppm, with an average of 436 ppm over 18 samples, in the Egyptian deposits of the Nile Valley (El Kammar *et al.*, 1979), 800 ppm in the Cambrian deposit of Karatau (in Baturin *et al.*, 1972), 90 – 300 ppm in the Hobso Gol Basin, in Mongolia (Illyin *et al.*, 1976), 620 – 860 ppm in the Georgina Basin in Australia (Cook, 1972). The highest concentrations have been found in the Soviet deposits of Maikop and Mangyshlak, where concentrations of bone debris reach an average of about 1% in rare earths (Kochenov and Zimovieff, in Altschuler, 1973); Baturin *et al.* (1972) even mention proportions of 4.7% in such facies.

Table 14 shows the general distribution of lanthanides, yttrium and scandium in 67 samples in the Soviet deposits, three composite samples from Florida and a sample from the Nile Valley.

The concentrations given in the first column of Table 12 are obviously exceptional and are due to the particular characteristics of bone debris. The results given by Semenov *et al.* (1962) for 53 Soviet phosphate samples of various origins, but not including this particular facies, show more modest concentrations, amounting to between 400 and 1000 ppm, with an average of 800 ppm. Studies available suggest that La, Ce and Nd are usually the most common lanthanides.

Chapter 5

Mechanisms of phosphate sedimentation and phosphatite accumulation

The presence of phosphatite deposits in a sedimentary basin requires an adequate source of phosphorus, as well as conditions favourable to the formation of apatite in the sediments and to the concentration and accumulation of the phosphate minerals formed in this way. Each of these factors will be discussed in turn.

1. POSSIBLE SOURCES OF PHOSPHORUS

Although geologists are generally agreed in regarding the apatite of magmatic (igneous) rocks as the primary source of phosphorus in sedimentary deposits, widely divergent views are held on the mechanisms by which such apatite is transformed into phosphatite. There are two main theories: that the sediments are directly fed with phosphorus from a continental source, or with dissolved minerals previously stored in the ocean depths.

1.1. Direct continental supply

The main advocate of a direct continental supply as the means by which phosphatite sediments are formed is Bushinski (1964). He quotes the example of the Volga, which discharges 6000 tonnes of phosphorus into the Caspian Sea every year in the form of mineral phosphate in solution, and twice as much again in an organic form. Even if only partly used in sedimentation, an addition of this magnitude seems sufficient to produce a very large phosphorite deposit within a few million years.

For a deposit to form in or near estuaries, very special conditions are necessary, which limit the generally enormous quantity of terrigenous material carried in suspension. Bushinski follows Strakhov (1962) in associating the formation of such deposits with dry climatic conditions, and suggests that a dry climate downstream and a humid climate upstream may increase the proportion of phosphatic material carried by a river, at the expense of the terrigenous fraction.

Soil development in a warm and humid climate, with its intensity and its 'filtering' function (Erhart, 1956), seems therefore to be a factor in increasing the supply of phosphate solutions to the basin, while reducing the quantity of detrital material carried. Such an environment has been used to explain the Eocene phosphate deposits of the Gao Basin in Mali (Radier, 1959) and the coastal sedimentary basin of the Togo–Benin region (Slansky, 1962). The study carried out by Westin and Brito (1969) also supports this idea and shows that in Venezuela the quantity of phosphorus remaining in the soil, i.e. not transported beyond the weathering profile, decreases rapidly as the intensity of weathering increases.

In fact, phosphorus does not behave in the same way in all soil profiles, and may be retained within the profile however intense the weathering. It occurs in a number of forms in soil, as shown by Walker (1972): some of the mineral solution emanating from the weathering of parent rock leaves the profile, some produces organic phosphates, while the remaining phosphate ions are associated with hydroxides of iron and aluminium. The latter fraction ('non-occluded phosphates', to use Walker's term) plays an active part in the evolution of the profile and in plant nourishment, until an external film of hydroxides, or a process of concretion, turns it into an immobilized phosphate fraction ('occluded phosphate') which is practically inactive in the profile, and which accumulates with insoluble residues of the parent rock. Figure 7a shows the interrelationships between these phases, suggesting the possibility of a slight external contribution of phosphate from rainwater. Figure 7b shows a possible way in which these fractions evolve as soil erosion progresses.

The proportion of phosphorus leaving the profile, which is shown as being considerable in Fig. 7b, may be far smaller, particularly on basaltic parent rock (Williams and Walker, 1969). According to Dabin (personal communication, 1979) it is only the initial phase of weathering that causes any appreciable loss of phosphorus; when most of the calcium is eliminated and the pH value falls below 5.5, almost all the phosphorus is immobilized in the form of iron and aluminium, which are only capable of migrating – and then generally only over short distances – under the effect of considerable concentrations of mobile organic acids (fulvic acids).

The alumimium phosphorites found in large quantities on the surface of various phosphatite deposits (Senegal, Florida, Brazil, Christmas Island, etc.) also show that it is possible for phosphorus to accumulate in the weathering profiles.

The extent of possible variation in soil development, even when it is rapid, is not as serious a limitation on its ability to act as a supply of phosphate to a basin as the dryness of the climate at the time of phosphate sedimentation: recent investigations have tended to confirm this. The palynological content of coastal deposits of the Gafsa Basin in Tunisia or of the Quseir region in Egypt (Fauconnier and Slansky, 1978), which is often very rich, is an indication,

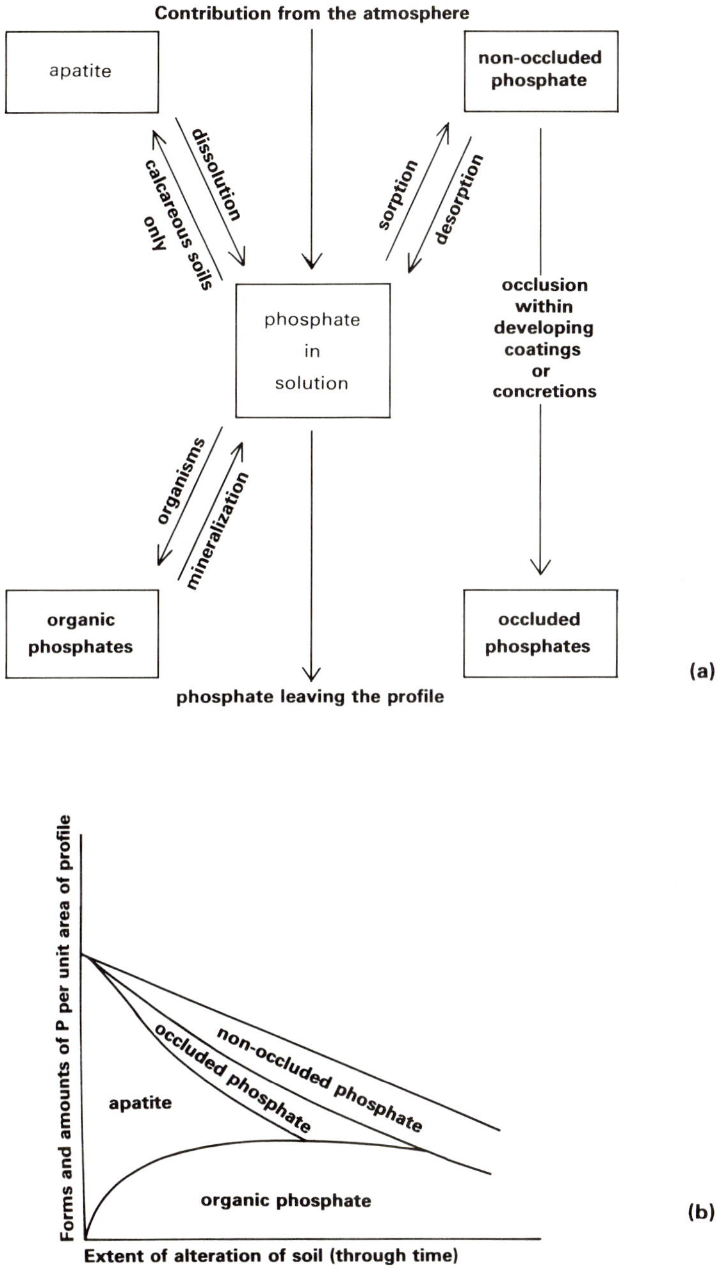

Fig. 7. *(a) Forms of phosphorus in soil. (b) Development of phosphorus forms as soil evolves* (from Walker, 1972).

among other things, of scarce vegetation on the nearby landmass; in Tunisia the aridity of the climate is, moreover, confirmed by the very low proportion of terrigenous sediments and the high proportion of evaporites (Sassi, 1974). In the Eocene sedimentation of the coastal basin of the Togo–Benin region, soil development is unlikely to have played as important a part.

If, therefore, the supply of phosphate to a sedimentary basin is directly attributed to a continental source, as in Pevear (1966) for deposits on the Atlantic coast of the U.S.A., and Keyser and Cook (1972) for deposits in Queensland, Australia, then it is Bushinski's view, with two climatic zones and with phosphorus drawn from weathering profiles far removed from the basin, that seems the most suitable.

It is difficult, however, to regard this view as generally valid, and numerous authors today prefer explanations based on the availability of phosphorus resources in the deep oceanic waters.

1.2. Supply from oceanic waters

The sea contains an average of 70–75 μg of phosphorus per litre, most commonly in one of four different forms, either in solution or suspension:

- in solution: phosphate ions or soluble organic compounds of phosphorus;

- in suspension: insoluble or adsorbed phosphates, or organic compounds.

The proportion of phosphate ions in solution increases with depth, and these ions represent practically the whole of the phosphorus from depths of a few hundred metres or more. HPO_4^{2-} ions, which predominate in the pH range between 6.1 and 10 (Gulbrandsen, 1969), are the most common; they are associated with small quantities of $H_2PO_4^-$ ions and still smaller amounts of PO_4^{3-} ions.

Oceanological measurements show that the quantity of phosphate in solution in the sea increases with depth, up to about 1000 m. There is a parallel drop in pH and temperature. Figure 8 gives examples of the distribution of phosphorus as a function of depth. This diagram is based on Gulbrandsen and Roberson (1973), who use data provided by Bruneau, Jerlov and Koczy (1953) for the North Pacific and the Indian Ocean, and by Worthington (1958) for the North Atlantic.

The Pacific Ocean contains the highest average concentrations of phos-

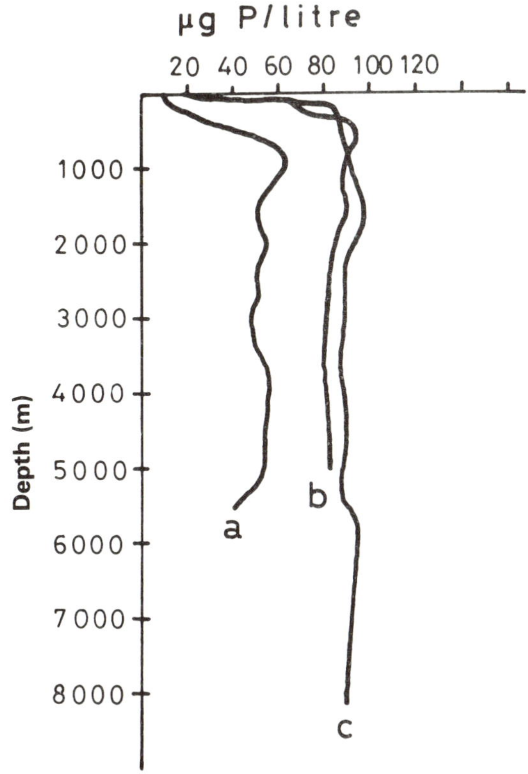

Fig. 8. *Variation in dissolved phosphorus content with depth: a — North Atlantic, b — Indian Ocean, c — North Pacific.*

phorus, while the Atlantic Ocean generally contains only half as much, except in the South Atlantic where there are levels comparable with those of the Indian Ocean. In the Mediterranean and the Red Sea, concentrations are generally below 30 or even 20 µg/litre, while in the Black Sea they reach more than 200 µg/litre at depth.

The general increase in the concentration of phosphate with oceanic depth makes it possible to consider the possibility of a feed mechanism for phosphate sedimentation from these deep oceanic resources; when the deep cold waters, moved by the upwelling currents, mix with warmer surface waters at the edges of basins, a higher than normal concentration of HPO_4^{2-} ions builds up in the basin, making a phosphate deposit more likely.

It was Kazakov (1937) who first proposed such a mechanism. The hypothesis was taken up by McKelvey *et al.* (1953) and various other authors. It has proved particularly fruitful because of its applicability to most of the world's deposits, including the present or Recent phosphate deposits of the sea floor, all of which are in zones with strong upwelling currents. In Baja California, d'Anglejan (1967) even estimated the annual quantity of phosphate in solution contributed at present by these currents over 225 km of continental shelf to be a minimum of 7×10^{15} tonnes. This considerable input is obviously only partially used in sedimentation (3×10^3 tonnes per year according to d'Anglejan), since it exceeds the oceanic resources, estimated at approximately 2.2×10^{11} tonnes of P_2O_5. If we adopt the estimate of 4.5 million tonnes of P_2O_5 generally given for the annual increment of new material to this resource, we may assume that it is wholly renewed about every 50 000 years. This 'period of residence' of phosphorus in the sea is therefore comparable with that of silica or barium (Gulbrandsen and Roberson, 1973).

2. FORMATION OF PHOSPHATE MINERALS FROM MARINE SOLUTIONS

According to Kazakov's hypothesis, phosphate deposits are formed by direct chemical precipitation due to supersaturation when cold water, enriched in phosphorus, rises from the deep ocean and meets the higher temperature, higher pH value, and lower CO_2 pressure of the upper waters; this precipitation may occur from depths of 200 m and above, stopping at 50 m in the photosynthesis zone, where the phosphate content is mainly consumed by phytoplankton.

Adopted and possibly modified by various authors, particularly McKelvey *et al.* (1953) for the Phosphoria Formation of the Rocky Mountains, this mechanism of direct chemical precipitation in the open sea is, in general, no longer accepted, for the following reasons:

1. If studies by Smirnov *et al.* (1962) and Roberson (1966) do seem to indicate that increased pH values assist apatite precipitation, Gulbrandsen (1969) considers that such an increase, at the pH values found at shallow depths, will be a far greater stimulus for the precipitation of $CaCO_3$. An experiment conducted by Gulbrandsen and Cremer (1970) using seawater resulted in a precipitate containing 43.0% CaO, 39.99% CO_2 and only 0.064% P_2O_5.

2. Martens and Harriss (1969) showed experimentally that Mg ions tend to inhibit apatite precipitation in the sea, and that at pH values of 7.5–8 this precipitation could only take place with a Ca/Mg ionic ratio greater than or equal to 5.2, which is well above the value generally found in seawater, usually around 0.20. Nathan and Lucas (1976) confirm this lower limit of 5.2 for the Ca/Mg ratio for seawater with a pH value of 7.8–8.1.

The possible supersaturation of seawater in relation to apatite is also disputed by Atlas and Pytkowicz (1977), in view of the particular properties of the surface of this mineral. It also appears doubtful in the light of experiments carried out by Savenko (1978) in connection with the precipitation of calcium phosphates. These arguments suggest that phosphate minerals do not form in the open sea, and most modern hypotheses are based on the development of depositional environments in which phosphate enrichment is brought about by marine organisms.

The idea that organic activity plays a part in the creation of phosphate deposits is not new: Murray and Renard (in Cayeux, 1950, Vol. III, pp. 903–904), although not the first in the field, were already working on it in 1891, after an examination of samples dredged up by the *Challenger*. Later, however, the lack of firm quantitive data on the possibility that phosphorus might be provided by organic means, and the apparent rarity of fossils in certain large deposits, made it necessary to seek alternative explanations for the creation of phosphatite deposits, including that put forward by Kazakov.

Recent oceanological data on the development of marine populations and on the present or Recent phophate deposits of the sea floor have made it possible to again take up the question of organic agencies with far more confidence.

2.1. Concentration of phosphorus by marine organisms

Numerous marine organisms concentrate phosphorus, as mentioned in Chapter 1. Figure 9, drawn from data provided by Slansky and Fauconnier (1973), gives details of the often very high levels of algae present in the seas and shows, for example, that the Phaeophyta, such as Laminaria or *Fucus vesiculosus,* contain 3 – 5% P_2O_5 on average (in their mineral fraction): comparable or even higher concentrations are found in the Rhodophyta.

The CaO/P_2O_5 ratio may sometimes be close to that found in sedimentary apatites, but is frequently far higher. These algae, moreover, concentrate calcium rather than magnesium (Fig. 10), since the MgO/CaO ratio is always far below its value of 3.7 in seawater.

Phytoplankton is of interest because of its frequent association with phosphate deposits. Diatoms are very abundant in sediments with phosphate nodules, which are found on the present sea floor, in the Miocene deposits of California and Peru, or the Lower Eocene deposits of Tunisia, etc. Dinoflagellates are also common in zones where phosphate production is still very high and they are plentiful in various older deposits (Fauconnier and Slansky, 1978; Fauconnier, 1979). The association of the Coccolithophorids with phosphate deposits (Lucas and Noel, 1957) is less certain at the moment: if these forms are abundant in deposits with a low phosphate content (e.g. the

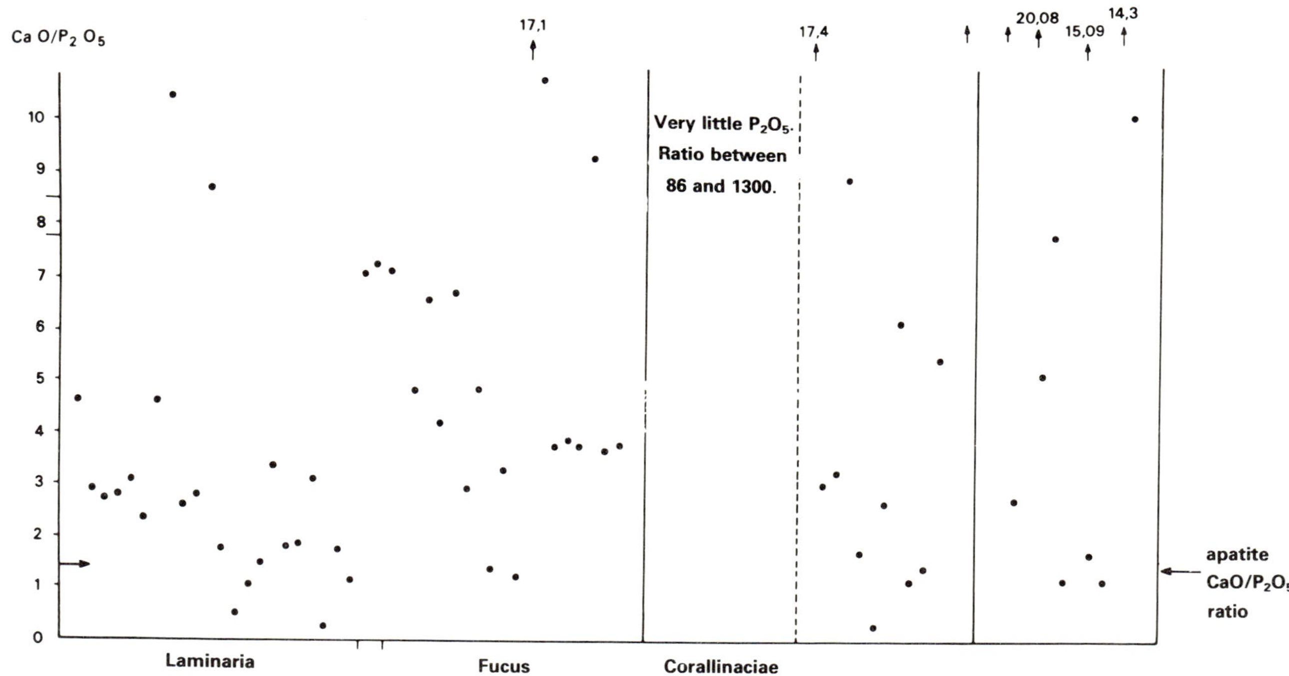

Fig. 9. *CaO and P_2O_5 contents of the mineral fraction of various algae, and the corresponding CaO/P_2O_5 ratios* (from Slansky and Fauconnier, 1973; and data published by Vinogradov, 1953).

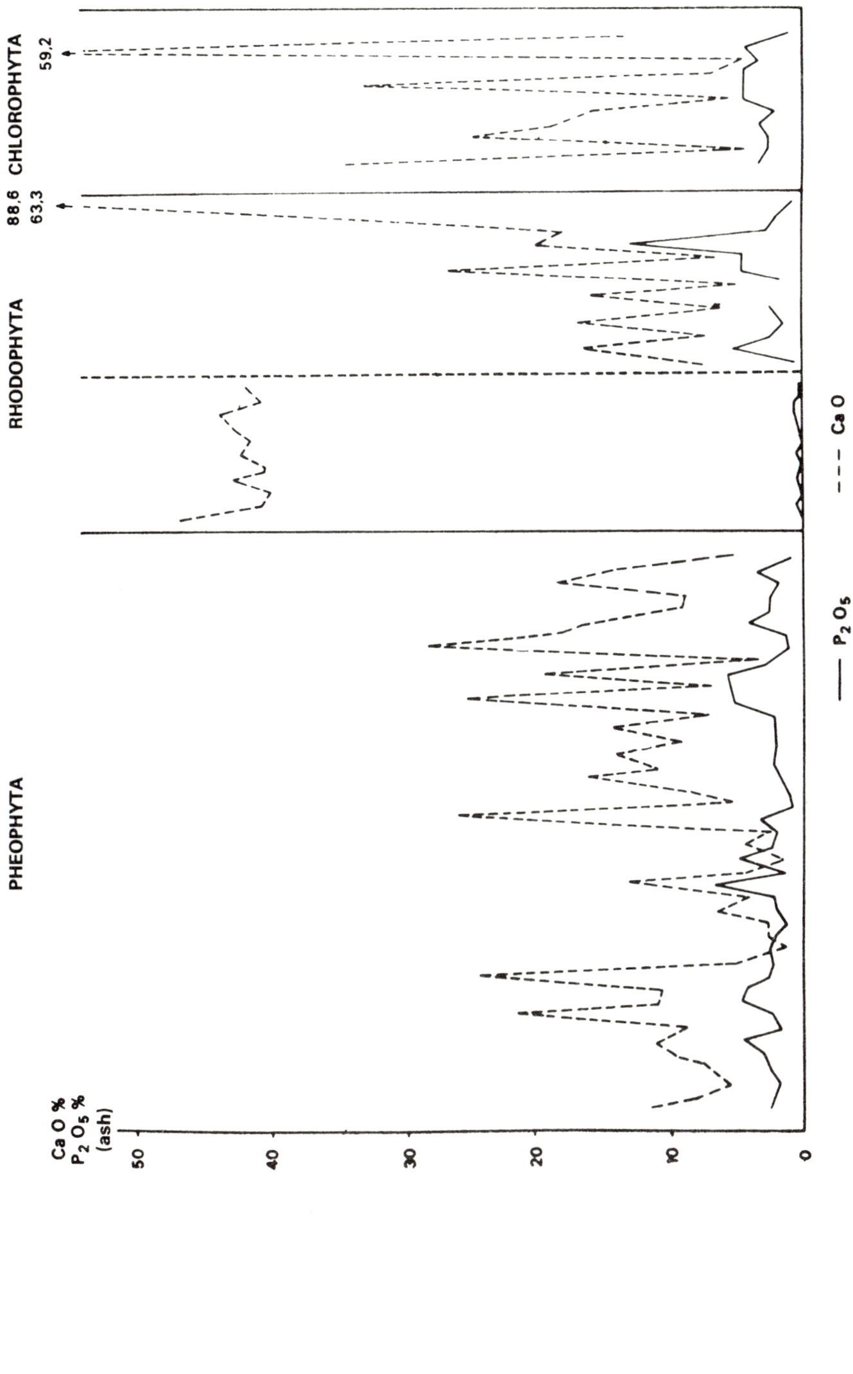

Fig. 9 Continued

phosphoric facies of the Albian and the Senonian of the Paris Basin), they are rarer in richer deposits.

According to Pérès and Devèze (1964), the diatoms and the dinoflagellates have the following partial chemical composition, expressed in oxides per 100 g of organic carbon.

	Diatoms	Dinoflagellates
C	100.00	100.00
P_2O_5	6.18	3.90
$F3_2O_3$	13.70	4.80
CaO	17.50	3.80
SiO_2	119.20	14.10

It is diatoms that have the highest P_2O_5 content. If we only consider the mineral fraction, however, we find that phosphorus is more abundant in dinoflagellates (14.6% of the total P_2O_5 + Fe_2O_3 + CaO + SiO_2) than in diatoms (3.9% of the same total), where the silica of the frustules tends to predominate. Where Coccolithophorids are concerned, it is the $CaCO_3$ content that is dominant and responsible for the marked reduction in the phosphorus concentration in the mineral fraction.

The role played by phytoplankton in phosphate sedimentation has long been underestimated, because of the small size of these organisms. For example, the volume occupied by the dinoflagellate *Gymnodinium brevis* is estimated at 7.5 $\times 10^{-9}$ cm^3 by Pérès and Devèze (1964, p. 267), with a phosphorus content of 0.48×10^{-10} g per individual specimen. This infinitesimal content becomes far more significant if account is taken of the 'red tide' phenomenon, which is liable to concentrate 1–100 million dinoflagellates per litre of seawater; for 10 million individual specimens the phosphorus content of seawater can therefore reach some 480 μg/litre, i.e. 5 – 6 times more than the content of the deep waters of the Pacific (cf. Fig. 8). If the same calculation was carried out with the figures quoted by Charles (1953, p. 173) for the peridinians, one would arrive at phosphorus concentrations of several grams per litre!

Another estimate (Riley *et al.*, 1949) suggests that phytoplankton have an annual organic carbon production of 475 t/km^2 in highly productive oceanic regions, i.e. an annual phosphorus production of 21 t/km^2, given that the carbon/phosphorus ratio for these organisms is about 50:1.

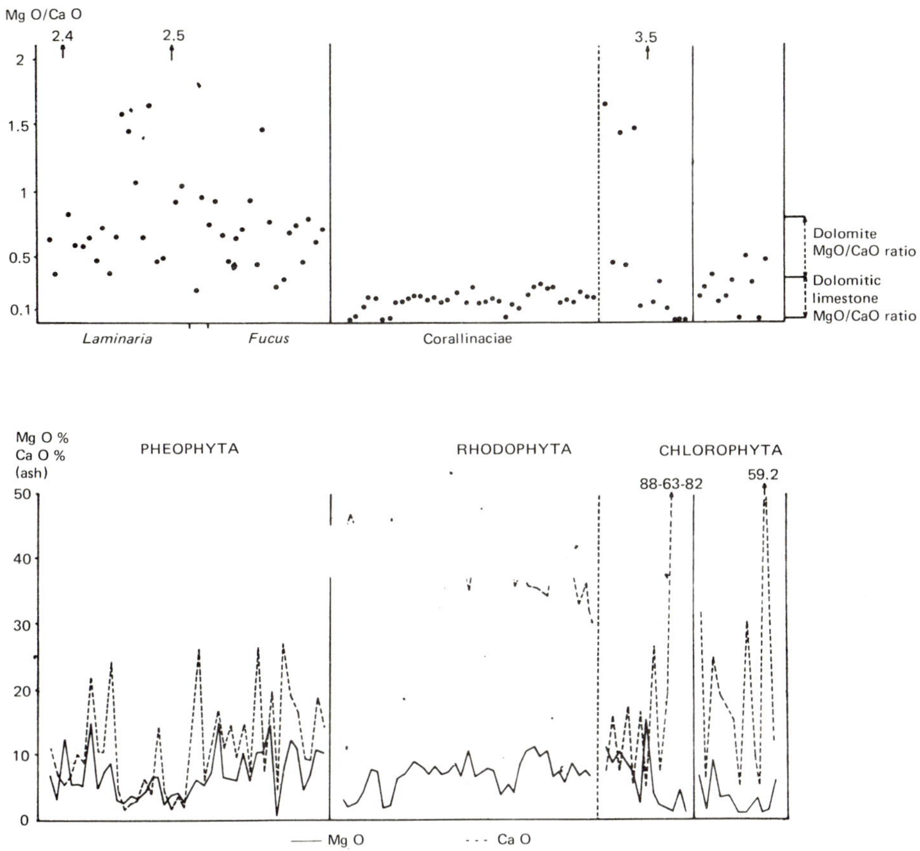

Fig. 10. *MgO and CaO contents of the mineral fraction of various algae, and the corresponding MgO/CaO ratios* (from Slansky and Fauconnier, 1973; and data published by Vinogradov, 1953).

2.2. Environment for the formation of phosphate minerals

In most cases, the phosphorus concentrated by such organisms is returned to solution in the sea and dispersed immediately after the organism's death. Rapid dispersion is only avoided if the basin has a 'trap' in which the organic matter can accumulate (Slansky and Fauconnier, 1973). It is within such an accumulated organic deposit that, given favourable conditions, sedimentary apatites will be able to form. The diatomaceous mud of the western continental margin of southern Africa, in which phosphate concretions are still being formed, provides a good illustration of the characteristics of such a deposit.

The diatomaceous muds of this region are known primarily as a result of the work of Baturin, in his numerous publications since 1969. According to Baturin (1971), these muds have the consistency of a semifluid gel containing 80 – 90% water. The dry residue comprises 50 – 85% silica emanating from the diatom frustules, 10 – 15% organic matter, and the remainder consisting of fine terrigenous material, embryonic gastropod shells, teeth and scales of fish, coproliths and phosphate nodules; the P_2O_5 content of 10 specimens taken at depths of between 75 and 128 m ranged from 0.32 to 1.35%, with an average of 0.81%.

Phosphorus concentrations in solution in interstitial water (Baturin, 1972) range from 118 to 2520 μg/litre in fresh samples; there is a fairly satisfactory correlation with the organic carbon content, which may reach 15%. After several months' storage at room temperature, the phosphorus content of the interstitial solution increases considerably and may exceed 8000 μg/litre.

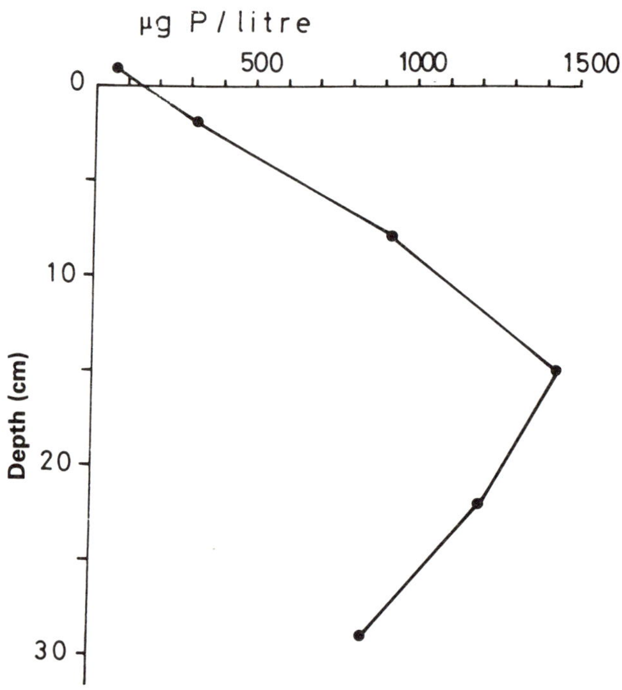

Fig. 11. *Phosphorus content of interstitial water in a core sample taken at a depth of 185 m on the Peruvian continental shelf* (from Sholkovitz, in Burnett, 1974.)

On the continental margin off Peru and Chile, the second region in which phosphate development has been shown to be occurring today from diatomaceous deposits, Burnett (1974, 1977) has obtained a comparable result for the phosphorus enrichment of interstitial water. Figure 11 shows that in a core sample taken at a depth of 185 m, the phosphorus content of interstitial water may reach about 1500 μg/litre some 15 cm below the water–sediment interface; the rise in the curve probably reflects the progressive dissolution of phosphorus in organic material, while the corresponding fall reflects apatite formation, beginning at a depth of about 15 cm.

The interstitial water of diatomaceous muds therefore has a phosphorus content far above that of seawater, which generally contains 80 μg/litre at most. They are heavily supersaturated in carbonate fluorapatite. In the case of the Namibian continental shelf, Price and Calvert (1978) show that the phosphorus content of interstitial water is up to 26 times higher than that of the seawater immediately above.

According to Baturin (1972), the pH value of interstitial water in the diatomaceous muds of the western continental margin of southern Africa ranges from

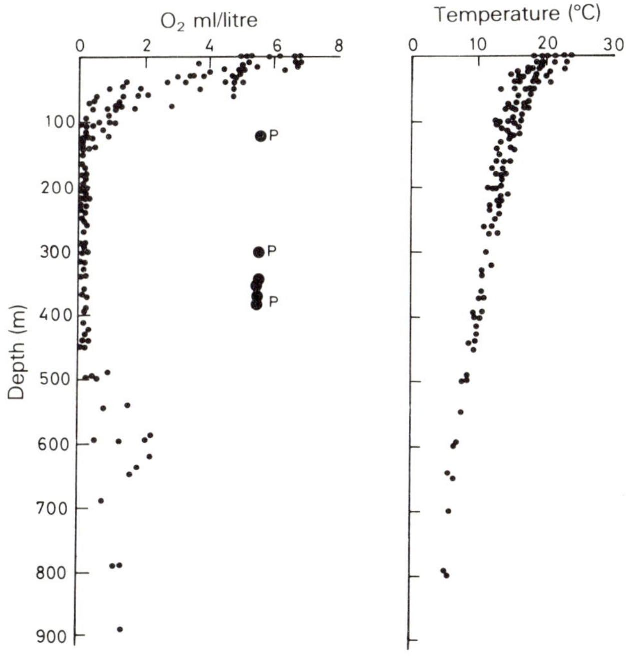

Fig. 12. *Dissolved oxygen and temperature variation as a function of depth, off the coasts of Peru and Chile; distribution of Holocene phosphate nodules in the same area* (from Burnett, 1974, p.138).

7.26 to 8.06, with a mean of 7.56 (based on 14 measurements) and it has a strongly negative Eh value (mean of –155 mV based on 14 measurements). The strongly reducing nature of muds containing phosphate nodules is associated with an oxygen deficiency in the overlying seawater, as is made clear by Burnett (1974) in his study of the phosphate deposits off Peru and Chile. Figure 12, which is taken from Burnett's thesis, illustrates this connection between the position of the O_2 minimum layer at depths ranging from 100 to 450 m, and the presence of phosphate nodules in the underlying organic deposit. More exact isotopic dating of these nodules (Burnett and Veeh, 1977) shows, however, that many of them are not recent, and that the most recent are not, in fact, in the O_2 minimum layer, but at its margins or even outside it (Burnett, 1980b). The environment in which the nodules form is not therefore so strongly reducing as might be assumed from the present position of the older nodules.

Belayouni's study of organic matter in the Tunisian phosphate deposits (1984) further shows that the environment is a stability field for humic acids of marine planktonic origin.

Humic compounds are organic substances which are not yet very well known, and are defined by the ways in which they can be extracted from the sediment or the soil that contains them. These compounds therefore include the fulvic acids, which are soluble both in bases and in acids, the humic acids, which are only soluble in bases, and humin, which is insoluble. What is known about the humic acids is that they are polymers of high molecular weight (from several hundred to several million), with acidic properties and enriched in oxygen by the presence of carboxylic and phenolic groups. Fulvic acids generally have lower molecular weights, while being richer in functional groups and in oxygen.

In the marine sedimentary basins, humic compounds form at an early stage of diagenesis, near the water–sediment interface, by condensation of monomers produced by enzymic degradation of deposits of dead plankton and necton. Fulvic acids apparently form first, followed by humic acids. Humic compounds are generally unstable and evolve more or less rapidly in the course of burial, by polymerization and loss of functional groups, eventually forming kerogen, an insoluble organic substance whose stability is only significantly affected by thermal degradation or surface weathering. This evolution is shown schematically in Fig. 13.

The stage in diagenesis that produces kerogen is more rapid in a reducing environment than in an oxidizing environment; if the environment is highly reducing from the moment the organic matter starts to evolve, the humic stage may in fact be absent (Tissot and Welte, 1978, in Belayouni, 1984). Conversely, if the environment is initially highly oxidizing, the organic matter is rapidly mineralized, and no humic compounds or kerogen are formed.

To be stable, humic acids therefore require a moderately oxidizing environ-

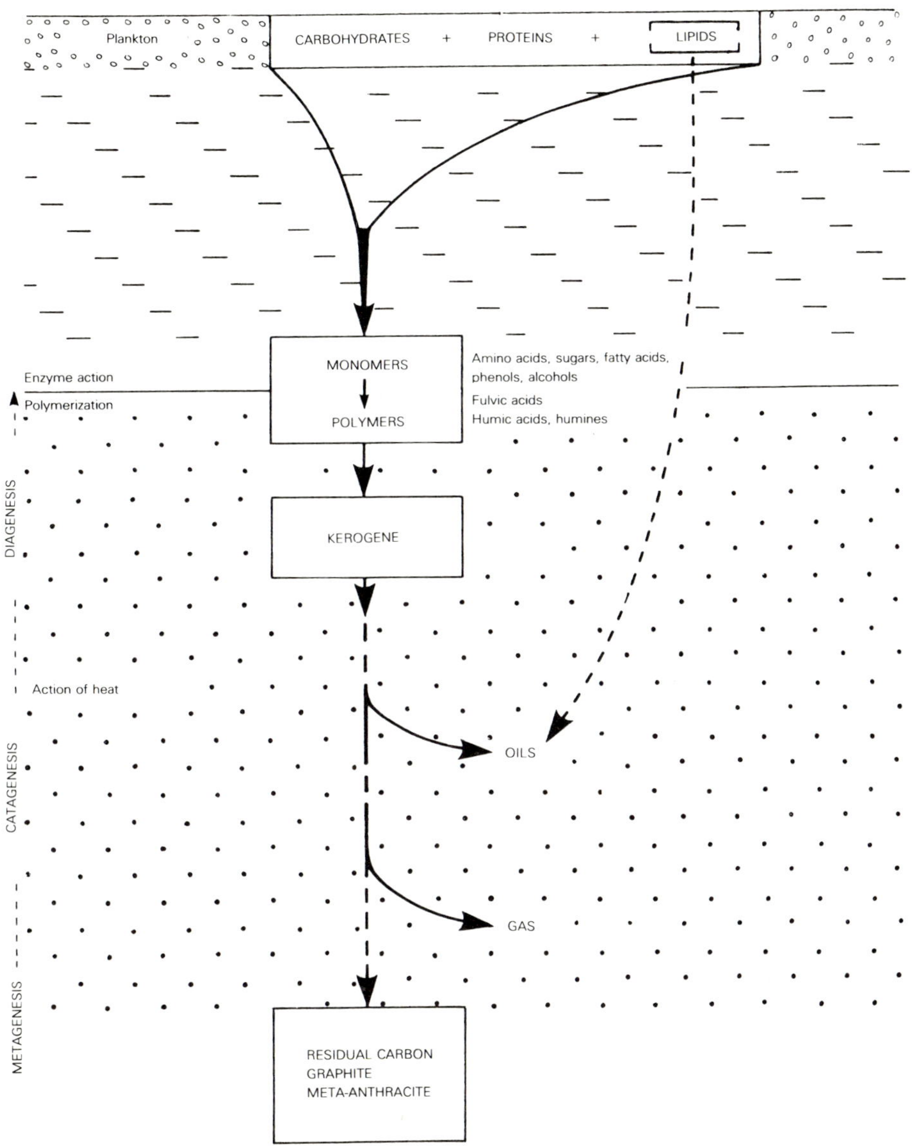

Fig. 13. *Evolution of marine organic matter* (adapted from Tissot and Welte, 1978).

ment, of the kind that may be found in the immediate vicinity of a reducing environment. In other words, the environment in which the new nodules form, near the margin of the marine O_2 minimum layer, also seems to be the most favourable for the stability of humic acids. Independent studies conducted by Poutanen and Morris (1983) have recently revealed over 60% humic compounds in the organic matter of the Peruvian continental margin.

The comparatively non-reducing or moderately oxidizing nature of the depositional environment in which the phosphate minerals form is confirmed by other studies of the organic content, to which we shall be referring later. It is interesting that as long ago as 1947, Teodorovitch suggested that the most favourable environment for phosphate development was near the boundary between the reducing and oxidizing environments, i.e. on the boundary itself ('neutral zone'), just above it ('slightly oxidizing zone') or just below it ('slightly reducing zone').

2.3. Formation of sedimentary apatites

The carbonate fluorapatites of phosphate deposits may be formed in one of three main ways: biogenesis, direct precipitation, or epigenesis.

2.3.1. Biogenesis

The tests of inarticulate brachiopods (Obulus, Lingula, etc.), whether complete or in the form of debris, found in large quantities in certain deposits (Ordovician in Estonia and in northern Argentina, and certain levels of the Permian in the Rocky Mountains, etc.), consist of francolite. The calcium phosphate content, therefore, depends directly on their accumulation and their favourable fossilization. The same applies to the debris of skeletons of fish or other marine or coastal vertebrates, which are also rich in phosphate; this debris, however, undergoes more rapid diagenetic change, mainly by conversion of the initial carbonate hydroxylapatite into carbonate fluorapatite. According to Baturin (1974a), the fluorine content increases from about 0.3% to over 3% in the case of fish. The main enrichment process is accompanied by a secondary increase in the proportion of iron, barium and rare radioactive elements.

The conodonts are also fossils consisting of francolite, but their initial state is still uncertain, and they are unlikely to have made any appreciable contribution to the formation of deposits.

Phosphatite coproliths are sometimes classified among biogenic phosphate particles. In fact, faecal pellets capable of fossilization have only very low phosphorus contents (0.2 – 0.4% P_2O_5, according to Moore, 1955). Consequently, unless mud-fed organisms from an environment already highly enriched in phosphate by some other means are present, the high phosphorus content of coproliths cannot have any direct biological origin.

2.3.2. Direct precipitation

Though difficult to envisage in an open marine environment, the direct precipitation of calcium phosphate appears far more plausible in organic deposits with interstitial waters rich in phosphorus. Burnett (1974, 1977) shows that this kind of precipitation takes place off the Peruvian and Chilean coasts, where very fine apatite crystals, in rods or grouped into 'rosettes', form on the surface of diatom frustules in the Recent organic deposits on the continental margin. Comparable crystals have been discovered by Baturin (1982) in specimens from the same region.

The same author (Baturin, 1969, 1974b) describes the formation of authingenic concretions on the southern African continental margin at depths of between 50 and 130m. The presence of concretions that have evolved in various ways reflects the different stages of phosphatization, from slightly phosphatic muds to compact concretions. Table 15 below gives the partial chemical composition of samples taken at station 2046 (22°40 S, 14°15.6 E) at a depth of 87 m.

According to Baturin, phosphatization is a diagenetic phenomenon which takes place in stages, the first being the deposition of gels of calcium phosphate from saturated interstitial solution. The structure of the initial mud is at first preserved, but later becomes blurred and disappears as dehydration and hardening takes place, during which the apatite crystallizes, more or less replacing non-phosphatic inclusions. Baturin and Shishkina (1973) also show that the fluorine content increases while the concretions evolve: on analysis, the

Table 15. *Stages in phosphatization off the Namibian coast* (from Baturin, 1974b).

Type of sample	P_2O_5 (%)	CO_2 (%)	Organic carbon (%)	SiO_2 (%) (opal)
Diatomaceous ooze	1.33	1.28	5.31	59.32
Slightly phosphatized ooze	3.56	1.07	3.97	38.00
Phosphatized ooze	19.51	4.32	2.25	14.24
Unconsolidated concretions	29.53	5.90	1.28	0.21
Dense granular concretions	30.31	6.21	0.97	traces
Compacted concretions	31.94	6.39	0.98	traces

F/P_2O_5 ratio is 0.07 in phosphatized diatom oozes, 0.08 in soft phosphate concretions, 0.09 in hardened concretions and 0.11 in phosphatic sandstones taken at a slightly greater depth of between 150 and 200 m.

Studies conducted by Price and Calvert (1978) on the Namibian continental shelf, however, do not point to the same form of evolutionary development, as non-consolidated phosphate nodules are already carbonate fluorapatites with an F/P_2O_5 ratio ranging from 0.10 to 0.14.

All this research appears to indicate two possibilities for the precipitation of calcium phosphate from the interstitial waters of organic muds: direct precipitation in the form of apatite (Burnett, 1974; Price and Calvert, 1978) or formation of an initial gel within which the apatite subsequently crystallizes (Baturin, 1969; Savenko, 1978).

Experiments carried out by Martens and Harriss (1970) provide an explanation for the existence of these alternative mechanisms. Working on the precipitation of calcium phosphate from artificial marine waters, they show that the addition of phosphate, or to the direct precipitation of crystallized apatite, depending on the Ca/Mg ratio. At a pH value of 7.5-8, this ratio must be greater than or equal to 5.2 for the apatite to precipitate. At lower pH values (Nathan and Lucas, 1976), the limit for this ratio is lower (1.2 at pH 7), although still higher than the value found in seawater, which is approxiamtely 0.2. Martens and Harriss have also demonstrated that the amorphous gel that precipitates when the Ca/Mg ratio is too low may rapidly evolve into apatite if placed in an environment containing few magnesium ions.

It seems reasonable, therefore, to suggest that this ratio plays a similar role in interstitial water, and governs the formation of the calcium phosphate in the same way.

Experiments carried out by Martens and Harriss (1969), and by Savenko (1978), however, show that the solubility of the amorphous gel of calcium phosphate is far higher than that of the apatite. According to Savenko, the solubility of the natural phosphates is of the order of 0.160 mg P/litre in seawater at a temperature between 18 and 23°C, whereas the amorphous gel will be in equlibrium with a solution whose concentration falls from 35 to 5 mg P/litre as the pH value increases from 7.3 to 8.3. At or above the usual concentrations of interstitial water, the direct precipitation of apatitic minerals is therefore apparently easier than the precipitation of a gel of calcium phosphate if the Ca/Mg ratio becomes sufficiently high.

In the organic muds of the phosphate-bearing basins, increases in this ratio may be due to a number of causes:

1. The organic content of the deposit may, by its very nature, assist in increasing the Ca/Mg ratio of the interstitial water. Figure 10 shows, for example, that an accumulation of algae generally has an MgO/CaO ratio lower than

that of seawater (3.7), i.e. a higher Ca/Mg ratio, which may exceed the limit of 5.2 discussed above, in the case of the Rhodophyta or Chlorophyta. Analysis of diatoms (reported by Vinogradov 1953, p. 151) also shows a Ca/Mg ratio that ranges from 0.7 to 2.7. Finally, this ratio may be increased by the dissolution of calcareous tests of the Forminifera, coccolithophorids, etc. which may be present in the deposit.

2. The formation of clay minerals rich in magnesium may also contribute to an increase in the proportion of calcium in solution. The formation of montmorillonite from the silica of the frustules of diatoms has been studied by various authors (Wollast and Debroeu, 1971; Chamley and Millot, 1972; Van Bennekom and Van der Gaast, 1976) in various contexts. A mechanism of this type can be applied to the Lower Palaeocene–Eocene deposit of the Gafsa Basin in Tunisia, where montmorillonite is the main

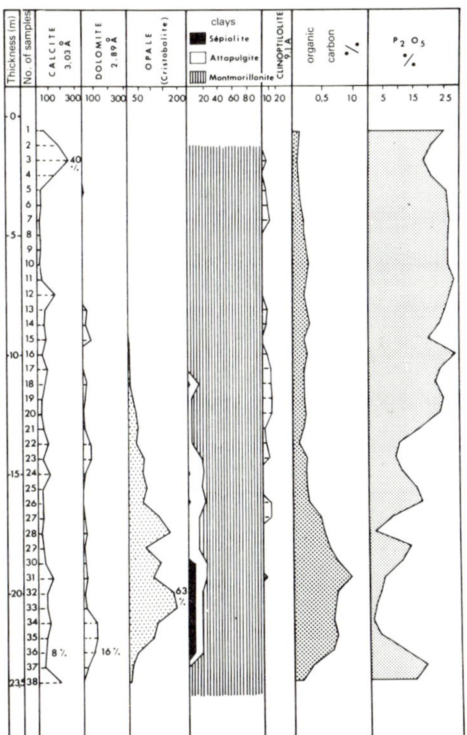

Fig. 14. *Relations between organic carbon, opal and clay minerals in the Palaeocene–Lower Eocene phosphatic sequence of the Gafsa Basin* (from Sassi, 1974). *Samples were taken from borehole SM 10 at M'rata. Variations in calcite, dolomite and opal contents were calculated from the results of X-ray diffraction measurements and chemical tests.*

clay mineral of a phosphate sequence originating from a deposit rich in siliceous organisms such as diatoms and radiolaria, or organisms likely to produce silica, such as the dinoflagellates; Fig. 14, based on data from Sassi (1974), also shows that the occurrence of attapulgite and sepiolite, the clay minerals richest in silicon and magnesium, is associated with the presence of large quantities of opal, whose biogenic nature is confirmed by a satisfactory correlation with the organic carbon content.

In the Santa Barbara Basin, near the Californian coast, Sholkovitz (1973) explains the reduction in the magnesium content of the interstitial water of deposits in the O_2 minimum layer at the bottom of the basin, by a further absorption of Mg^{2+} ions by clay minerals as a result of the loss, to form new sulphides of the Fe^{3+} ions associated with these minerals (on the surface or interstratified, but not in the lattice).

3. At last, the restricted nature of the environment in which the phosphate series form, favours the use of magnesium in interstitial water for the formation of dolomite, a mineral far more abundantly associated with phosphate sediments than used to be thought. Dolomite, for example, is frequent in the phosphate sequences of the Palaeocene and Lower Eocene of northern Africa (Boujo, 1976; Ranchin, 1963; Sassi, 1974).

The concentrations of phosphorus in interstitial water and the variations in the relative proportion of calcium and magnesium make it plausible to suggest that direct precipitation of apatitic minerals takes place in the course of diagenesis, within organic muds in low oxygen environments. The precipitation of amorphous gels of calcium phosphate, on the other hand, seems likely, given the average concentrations found in the interstitial solutions of modern diatom oozes; the considerable increases in concentration of these solutions (Baturin, 1972) nevertheless suggests that these may be 'privileged', partially isolated zones within the muds, in which there are sufficient quantities of phosphorus for gels to form. Furthermore, Berner (in Burnett, 1974, p. 140) reports far higher proportions of interstitial phosphorus, up to 1000 ppm, than those reported by Baturin.

The formation of sedimentary carbonate fluorapatites also depends, however, on the concentration of calcium, fluorine and carbonate ions in the environment.

Large quantities of calcium are provided by organisms: Fig. 9, for example, shows that algae often have a CaO/P_2O_5 ratio higher than that of apatite; the same applies to phytoplankton with diatoms. Baturin and Shishkina (1973) have also shown that in the producing zones of the southern African and Chilean continental margins, fluorine content increases from 1.3 ppm in the surface water to 1.8 ppm at depth, reaching 11 ppm in interstitial water.

At the same time, the F/P_2O_5 ratio falls from 14.5 to 8.7, or even 2.18, but always remains well above its value in francolite, approximately 0.1. Finally, carbonate ions, whose role in the precipitation of carbonate apatites is stressed by McConnell (1965), are produced by the decomposition of organic matter: according to Romankevich and Baturin (1972), the phosphatization of the diatomaceous mud of the western continental margin of southern Africa is accompanied by partial oxidation of organic carbon; 17–38% of the CO_2 produced in this way is then integrated into the phosphate minerals.

2.3.3. Epigenesis

Observations of phosphate grains originating from the replacement of previously calcareous bioclasts or lithoclasts have led various authors, such as Cayeux (1950), Lucas *et al.* (1977), to conclude that the epigenesis of carbonates plays an important or even essential role in phosphate sedimentation. Ames (1959) demonstrated experimentally that very slight concentrations of phosphate in solution (0.1 ppm PO_4^{3-}) are enough to trigger this kind of replacement, which he therefore regards as the main mechanism for the formation of marine phosphate deposits.

Despite the low proportions of phosphorus shown to be sufficient by Ames, epigenesis does not remove all the obstacles to the formation of apatite in an open marine environment, particularly those associated with the Ca/Mg ratio. In interstitial water, on the other hand, whose composition is different and in which ionic diffusion is slower, the fact that calcite passes into solution locally prior to crystallization of apatite during substitution, may make it easier to reach the necessary Ca/Mg value; this mechanism supplies not only calcium, but also the carbonate ions that francolite requires. Epigenesis is therefore a plausible mechanism in interstitial water.

Of the three modes of formation of sedimentary apatites discussed, direct precipitation and epigenesis seem to be the most common, though the relative frequency of each is not certain. Biogenesis, i.e. the direct biological formation of phosphate minerals, seems less frequent, although it may be the principal mechanism locally; however, this may be revised when the role of bacteria is more certain.

2.4. Factors triggering the formation of sedimentary apatite

Despite their ability to concentrate phosphorus, calcium and fluorine and to supply the necessary carbonate ions, not all phytoplankton oozes develop into phosphatic strata. Development in this direction depends on various other conditions related either to the basin's supply, i.e. to a supply of material particularly suitable to phosphate sedimentation or to the suitability of the depositional environment.

2.4.1. Suitable supply

Various authors have studied the possibility that volcanism may play an active role, directly or indirectly, in phosphate sedimentation. For example, Mansfield (1940) suggested that volcanic fluorine favours the formation of phosphate minerals, while Bidaut (1953) uses phosphorus fluorides from submarine fumaroles to explain the formation of the Dinantian deposits in the French Pyrenees. Kharin (1974) suggests that alkaline exhalations are responsible for the 10.3% of P_2O_5 in the siltites and clays found at present at a depth of 1000 m on the Jan Mayen submarine ridge to the north-east of Iceland; more generally, he considers that periods of intense volcanism cause an increase in the phosphorus content of seawater, and therefore stimulate the development of plankton which, in turn, will provide phosphorus for sedimentation. According to Lowe (1971), who studied the Miocene deposit of Upper Sespe Creek in California, acid volcanism may have a double effect: because of the ash produced and its alteration to bentonite, and as a result of reducing the magnesium content of seawater and increasing its pH value, the crystallization of carbonate fluorapatite is stimulated.

These different hypotheses draw support from the observation that a number of phosphate deposits are closely associated with volcanic phenomena. This is the case, for example, of the Mexican Miocene deposit of La Paz, Baja California, where the ore is difficult to enrich, notably because of the volcanic fraction it contains. The same applies to the Upper Cretaceous in Colombia, contemporary with a period of intense volcanism, the deposits of the Lower Palaeocene–Eocene of the Gafsa Basin, Tunisia, where Clocchiati and Sassi (1972) found evidence of several volcanic periods, and the phosphate levels of the Cambrian on the northern slope of the Montagne Noire in France, associated with acid volcanic flows and volcanoclastics (Donnot and Guérangé, 1978; Prian, 1980).

On the other hand, no trace of volcanic activity has been reported so far in many large deposits, such as those in the Rocky Mountains, Morocco, Togo, etc.

Another development in the supply of material which is likely to assist phosphate sedimentation is the 'red tide' phenomenon (or 'discoloured water' according to certain authors) which occurs periodically in regions of oceanic upwelling. This phenomenon, due to the particularly intense bloom of dinoflagellates, may – if it occurs frequently – increase the proportion of phosphorus in the mineral fraction of deposits, both from dinoflagellates and from the mass mortalities they cause among fish. This mass mortality is due in particular to a toxic substance produced by the dinaflagellates, known as saxitoxin (Pérès, 1976), which may affect various different points in the food chain, with damaging effects on many organisms, including even sea birds and possibly human beings.

It should be noted that a massive supply of fish remains to a deposit also has the effect of increasing the Ca/Mg ratio: an analysis of whole *Sardinia caerula*, cited by Vinogradov (1953, p. 499), puts this ratio at 7.7.

2.4.2. Suitability of the depositional environment

The formational environment of the calcium phosphate minerals suggested above is located near the water–sediment interface, within the organic deposit, and is therefore slightly reducing to moderately oxidizing in character. It generally lies in a basin between a reducing environment underlying anoxic seawater and an oxidizing environment associated with well-aerated waters.

The reducing environment, poor in benthic life and with little bioturbation is favourable to the formation of laminated facies rich in organic matter, in which siliceous plankton can accumulate and evolve slowly into cherts or clay minerals rich in silica, such as smectites, attapulgite or sepiolite, or even into silicates such as clinoptilolite (Hein *et al.*, 1978). These shales, moreover, are very similar to the 'pre-evaporitic laminated facies' described by Busson and Perthuisot (1977) and Busson (1978), which means that they are also likely to evolve, temporarily at least, into more clearly evaporitic deposits containing magnesium carbonate and gypsum. The oxidizing environment, on the other hand, is rich in benthic life and subject to considerable bioturbation, favouring the development of biogenic limestones and militating against the conservation of organic matter; the energy of the environment may also encourage detrital deposits.

Phosphate deposits therefore occupy an intermediate position between these two categories of facies, as shown in Fig. 15. Only the upper limit of the O_2 minimum layer has been shown – its lower reaches are not generally in the regions where ore deposits form. The upper limit may be called a pycnocline, given the varying degrees of confinement and therefore of density in the environments that it separates.

Figure 15 shows that different sediments such as attapulgite shales, phosphatites and biogenic limestones may form simultaneously in a basin, as a result quite simply of local variations in the depositional environment, i.e. independently of the nature of the supply. Similarly, these facies may alternate and replace each other along a vertical axis, as a function of variations in the position of the upper limit of the O_2 minimum layer during sedimentation.

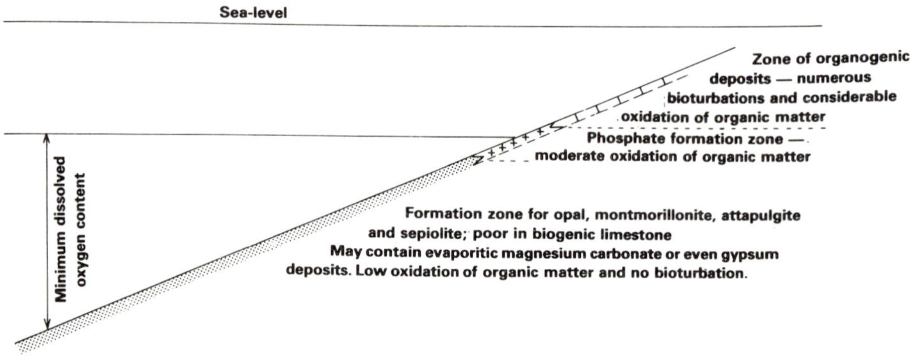

Fig. 15. *Schematic location of zone favourable for phosphate formation.*

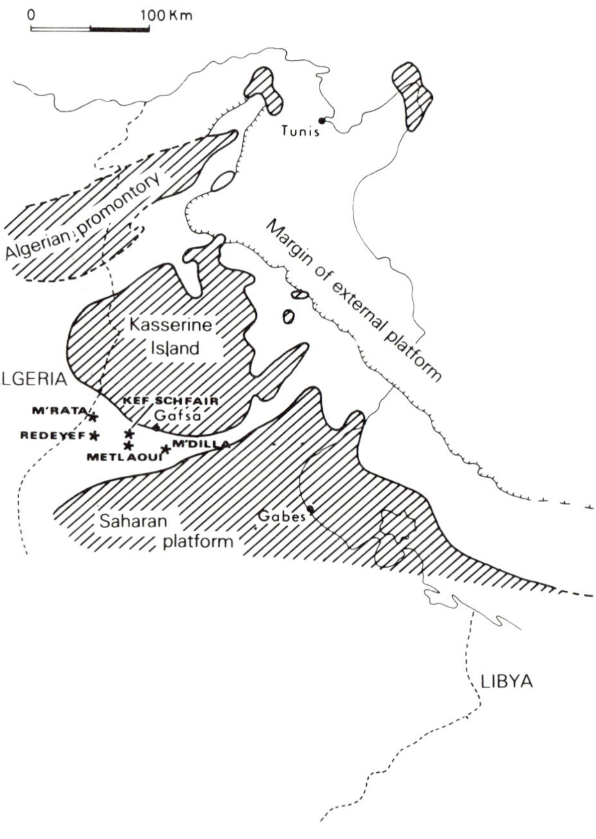

Fig. 16. *Location of boreholes in the Gafsa Basin.*

This dominant influence of the depositional environment on the nature of the deposit is in line with results obtained from the Lower Palaeocene–Eocene phosphate basin in Gafsa (Fauconnier and Slansky, 1980; Belayouni and Trichet, 1980). Figure 16 shows the positions of the boreholes studied, Fig. 17 summarizes some of the results obtained from samples collected at borehole S 234 at Kef Schfair, Tunisia, and Fig. 18 shows the relationships found between the proportions of P_2O_5 and the organic content of all the samples. Caution is needed in interpreting the results, as there are uncertainties associated with the methods used, and because of possible synsedimentary reworkings. We can, however, draw the following conclusions:

1. Throughout the phosphate sequence, organic matter is, in general, from the same marine planktonic origin. This is made clear by detailed study of the organic constituents, which reveals that dinoflagellate cysts, often present in great numbers, are far more common among the polymorphs than pollen grains.

2. The petroleum potential (the total hydrocarbon and hydrogen content, expressed in mg per g of rock) and the organic carbon content developed in parallel to each other, as does the concentration of microplankton, though here there are slight variations. The values of these parameters generally vary inversely with the P_2O_5 content; phosphate formation therefore corresponds to oxidation of organic matter. However, as shown in Fig. 18, high concentrations of P_2O_5 are often associated with considerable residues of organic carbon or microplankton, suggesting a moderate level of oxidation, as would be expected in the vicinity of the pycnocline.

3. The two specimens containing calcite (very abundant in both), shown in Fig. 17 (42.3 and 55.4 m), are poor both in organic content and in phosphate; the benthic fauna contained in them (pelecypods and foraminifera) confirm that they are formed in a well-oxygenated environment, above the pycnocline.

4. The sample taken at 49 m, on the other hand, which clearly contains dolomite, is rich in organic carbon and microplankton, with a high petroleum potential. It corresponds to a restricted environment clearly situated below the pycnocline, and its P_2O_5 content is very low.

It may also be seen from Fig. 18 that samples rich in opal are generally only slightly phosphatic , which might indicate that diatomaceous phytoplankton is less suitable for phosphate formation than phytoplankton containing mainly dinoflagellates: it may also be the case, however, that the frustules of diatoms have been better preserved from premature dissolution in the most reducing environments, i.e. in those least favourable to phosphate formation.

The influence of the environment on the nature of the deposit can be traced

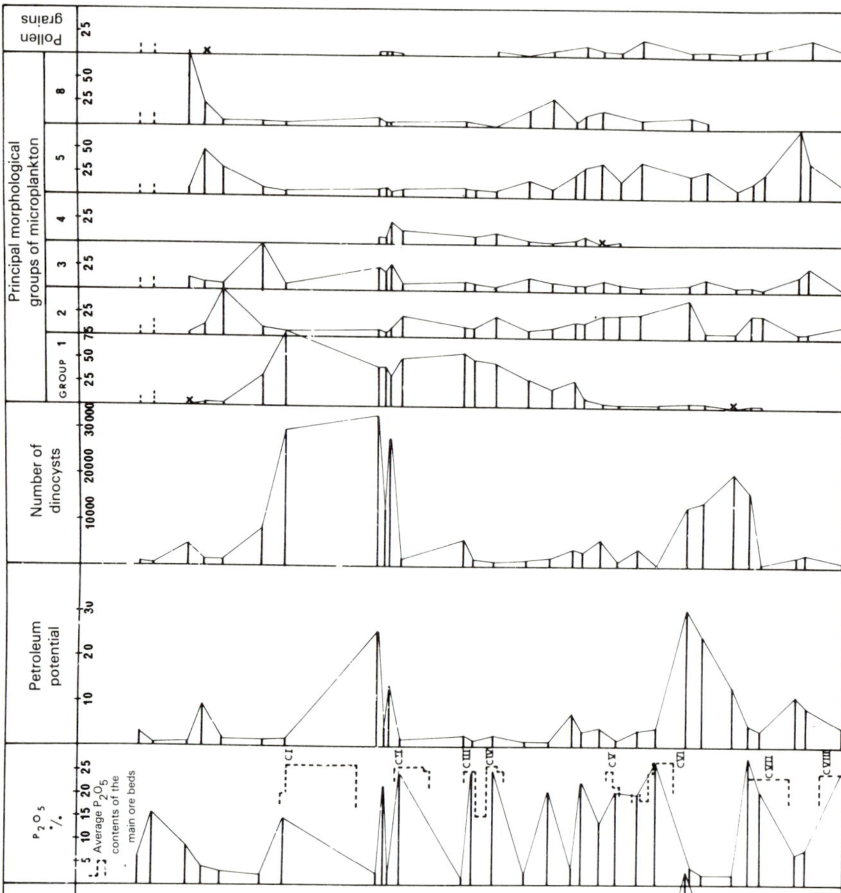

Fig. 17. Results obtained with samples collected from borehole S 234 at Kef Schfair.

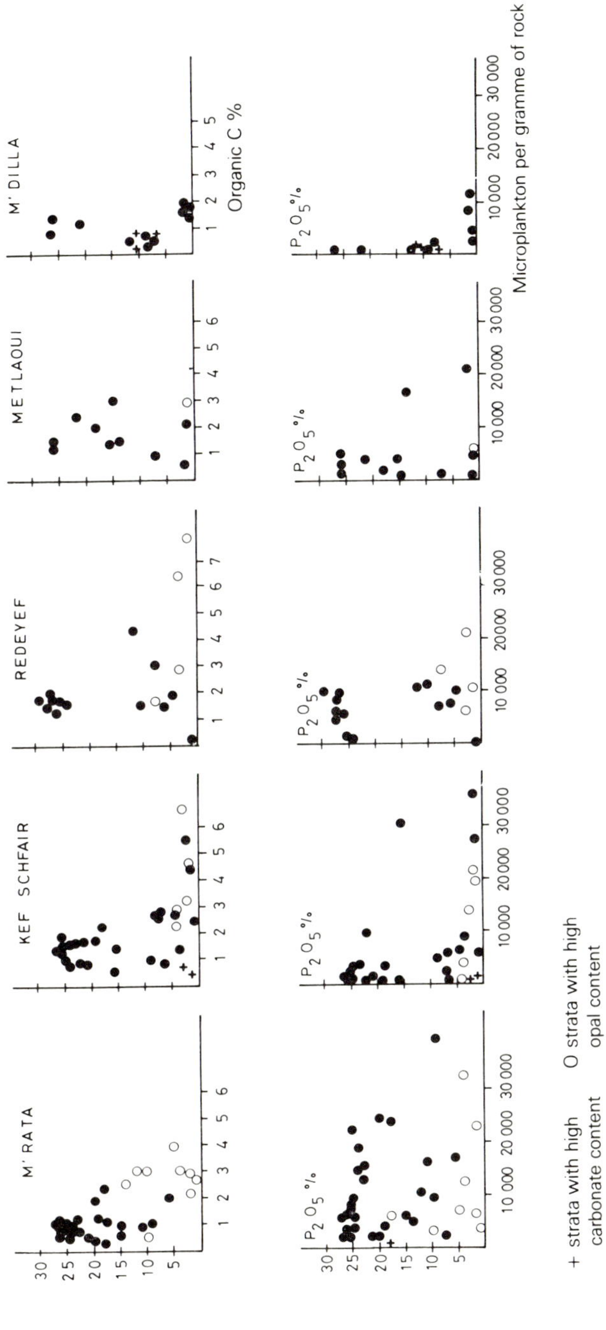

Fig. 18. *Relations between contents of P_2O_5 and organic matter, from boreholes in the Gafsa Basin.*

in the Eocene coastal basin in Benin, where phosphatic layers are associated with finely-laminated attapulgite-montmorillonite shale (Slansky *et al.*, 1959), typical of reducing depositional conditions, although they correspond to periods of more highly oxygenated water. Similarly, Romankevich and Baturin (1972) find that the modern phosphate nodules on the western continental margin of southern Africa contains less organic carbon (0.76 – 1.03%) than does diatomaceous mud (4.25 – 4.93%).

There appears to be no obstacle, therefore, to accepting as generally valid the idea that phosphate sedimentation is associated with the O_2 minimum layer of the sea and that the environment in which phosphates form is near this level, under conditions which are moderately reducing or oxidizing.

Why is it, however, that there is a particular tendency for carbonate fluorapatite to form in this environment? The results recently obtained by Belayouni (1984) provide an interesting answer. He shows that in the Eocene phosphate sequence of the Gafsa Basin there is an excellent correlation between the P_2O_5 content and humic acid content in the specimens studied (humic acids are represented by their organic carbon content); Fig. 19 gives an example of this correlation. Humic acids are primarily found inside phosphate particles; another study, carried out at the B.R.G.M. (Samb, 1983) shows, for example, that in coproliths of the Senegalese Eocene 88% of the organic carbon is represented by humic acids (96% for humic compounds as a whole).

Humic acids, however, are very unstable substances within the marine organic matter evolutionary cycle (cf. Fig. 13), so their continued presence in

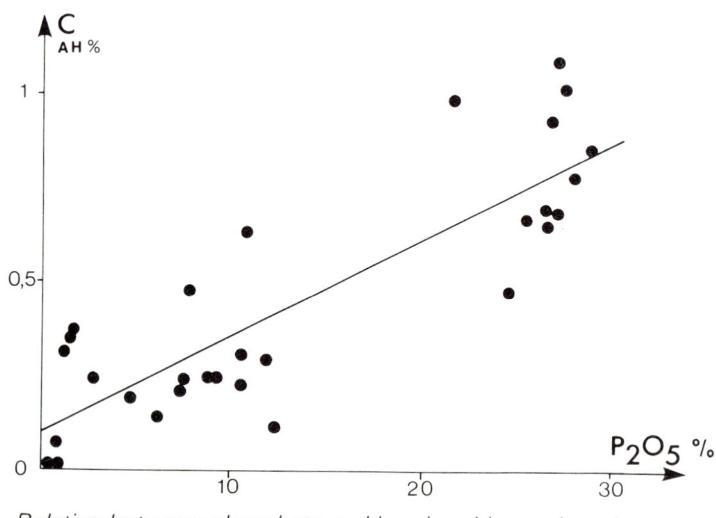

Fig. 19. *Relation between phosphate and humic acid organic carbon contents of 30 samples from the Gafsa Basin (Tunisia),* (from Belayouni, 1983) – *coefficient or correlation r = 0.85.*

such large quantities in old phosphate grains suggests that they have been trapped by phosphate minerals in the course of their formation. Belayouni explains this trapping by a chemical link using the calcium cation: humic acids–Ca–PO_4.

The close association between humic acids and phosphate also probably reflects trapping of phosphate dissolved in humic acids, which would itself be a factor stimulating the formation of apatite minerals. On this assumption, the most favourable conditions for phosphate deposits would be found in zones where the humic acids are least unstable and therefore most abundant, i.e. near the margin of the O_2 minimum layer of the sea.

This zone of highest humic content has the additional advantage of containing interstitial water enriched in phosphate. Nissenbaum (1979) has shown that after it has been deposited, planktonic organic matter releases most of its phosphorus during the transition from fulvic to humic acids.

There have still been very few analyses of humic acid concentrations in phosphorites. We do know, however, that these compounds are common in Cambrian ores of Australia (Sandstrom, 1982), the Upper Cretaceous in Israel (Amit and Bein, 1982), the Eocene of Morocco and Togo and the Miocene of North Carolina (Rachidi, 1983). Assumptions concerning the role played by humic acids in the Gafsa Basin do, therefore, seem plausible.

Accordingly, the results of recent studies on organic matter throw new light on the role of the environment in phosphate sedimentation. Nevertheless, the evidence is still only partial because the term 'humic acids' covers a large family of organic compounds of which we still know very little. Furthermore, other organic substances may also contribute to the formation of phosphate minerals. McConnell (1965), for example, emphasizes the possible effects of the enzyme carbonic anhydrase, which has been found in particular in the mantle of Lingula near its francolite test. He used this enzyme for the precipitation of carbonate apatite in an artificial environment. Other enzymes – phosphatases produced by algae – are capable of accelerating the extraction of phosphorus from organic matter (Jansson, 1977). Conversely, organic phosphonates may inhibit the growth of apatite crystals, as has been shown by Meyer and Nancollas (1973) for hydroxyapatite.

McConnell (1965) also stresses the probable role played by bacteria, which are capable of causing apatite precipitation. Riggs (1979b) too, attributes a predominant role to bacteria in producing the phosphate particles of the Florida ore.

It is obviously tempting to assign an important role to bacteria as they are naturally rich in phosphorus, containing 8% P_2O_5, dry weight (Bowen, 1966); phosphorus may be stored in bacterial cells, particularly those that have stopped subdividing, in the form of polyphosphate inclusions (Tardieux-

Roche, 1966) or even as hydroxyapatite (Ennever, 1963). Furthermore, very large numbers of bacteria are found on the sea floor of the continental margin off Peru, an area where phosphate sedimentation is taking place today (Froelich, 1983), and work carried out by Cayeux since 1936 (Cayeux, 1939, 1941, 1950), involving detailed petrographic examination under infra-red light, shows that bacteria are very common in the older phosphatites. Organic substances in phosphate rocks also contain clear evidence of bacterial action, such as hopanic derivatives (Belayouni, 1984).

Lucas and Prévôt (1984) also mention the possibility of an indirect role for bacteria in the formation of apatite, in that they contribute to fixing dissolved magnesium. In their experiments on apatite synthesis in seawater, they use organic phosphates (RNA) whose bacterial degradation leads to magnesium being trapped in a phosphate mineral, struvite ($NH_4MgPO_4.3H_2O$).

Although bacteria are undoubtedly very common in the depositional environment of sedimentary apatite, the role they play in apatite formation is, however, still uncertain.

2.5. Formation of phosphate particles

Certain phosphate deposits show strata, beds or lenses of microsphatite, often of secondary origin. In most cases, however, the phosphate in sedimentary rocks occurs in the form of clearly distinct particles the size of arenite or occasionally rudite. They are produced by five main mechanisms:

1. Reworking of pre-existing particles.
2. Epigenesis of pre-existing non-phosphatic particles.
3. Filling of cavities or porous media.
4. Phosphatization of faecal pellets.
5. Direct precipitation on a suitable surface.

The *reworking of pre-existing phosphate particles* is obvious in the case of bioclasts formed from test of inarticulated brachiopods that originate as francolite in composition, or skeletons of vertebrates that develop into francolite in the course of fossilization. In the Mio–Pliocene deposits of Florida, Riggs 91979a) extends this mode of formation to practically all the phosphate pellets, which are regarded as intraclasts formed by reworking of a pre-existing deposit of microsphatite, itself precipitated chemically or bacterially. On a smaller scale, Prian (1980) suggests a comparable origin for the mineralized fraction of certain granular phosphatic facies of the Cambrian of the northern slope of the Montagne Noire, believed to be derived from the reworking of undurated phosphatized surfaces of contemporaneous sediments. Near Udaipur, in India, some Precambrian phospharenites were also formed by a reworking of this kind, the initial facies being a stromatolitic microsphatite.

Reworking in these cases is primarily mechanical, and involves transport, sometimes over a considerable distance; but biological effects are never wholly absent. Certain organisms ('burrowers' or 'borers') may, for example, contribute to breaking up the original deposits, while others are involved in shaping grains – Riggs, for example, observes that in Florida certain categories of grain acquired their rounded shapes before transport, as a result of the agglomeration of primary phosphatic mud by Cyanophyta or bacteria, or by conversion to faecal pellets by worms or crustaceans.

Reworking may also be solely or mainly biological. This is particularly the case with fish, whose remains are highly sensitive to bacterial attack; the very unsatisfactory fossilization of their skeletons may therefore be totally independent of any movement at the sea floor, as is also frequently revealed by the well-preserved teeth obtained. Certain parts of the skeleton may be converted into pellets by 'phosphomicritization' due to the action of Schizomycophyta, as described by Soudry (1979), then Soudry and Nathan (1980) in the phosphate deposits of the Negev; when conversion is complete the origin of the pellets is no longer identifiable. As a comparable phenomenon can be observed in a specimen of the Palaeocene of the Gafsa Basin, there is good reason to accept that pellets derived from these bioclasts had the same rounded shape from the start (Plate 11, Photograph 48).

Substitution by calcium phosphate may affect all types of particles, including wood fragments, as in the Albian of the eastern Paris Basin (Hilly and Haguenauer, 1979, p. 37). The most frequent form of epigenesis, however, is associated with calcareous lithoclasts: more or less rounded fragments of micrite, debris of tests of pelocypods, echinoderms, etc. Organic activity is certainly involved in these substitutions. Photographs 47a and 47b (Plate 10), for example, show the start of phosphatization of an oyster shell, a process comparable with the algal micritizations of mollusc tests described by Bathurst (1971).

Phosphate grains formed solely by epigenesis, however, are not, apparently, very abundant. Particles combining different modes of formation are more frequent: the epigenesis of foraminiferal tests is, for example, often associated with a generally earlier stage involving a phosphate filling of the chambers, and the formation of an external pellicle of carbonate apatite. Champetier and Joussemet (1979b) found that these foraminifera include Nubecularia which produce pseudo-oolithic grains; they also discovered oncoliths whose micritic lamellae have been substituted by phosphate.

The phosphate pellets formed by filling the chambers of foraminifera are fairly frequent, e.g. in the Eocene ores of Taïba (Senegal), Guinea Bissau and

Plate 10

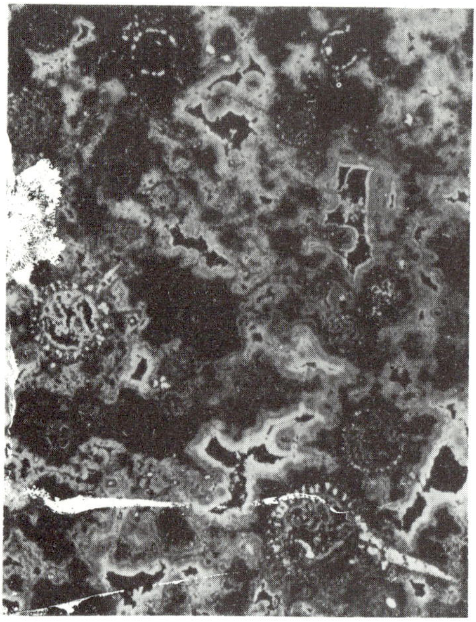

45a, LN 0.1 mm

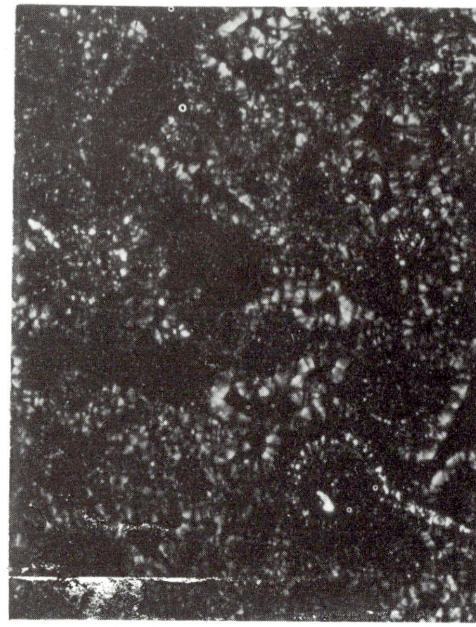

45b, LP 0.1 mm

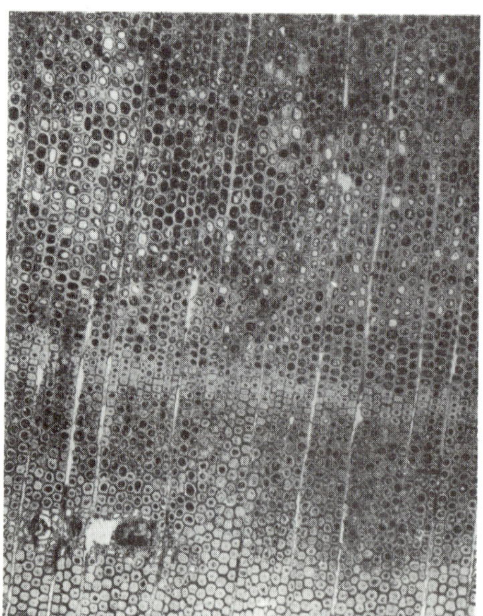

46a, LN 1 mm

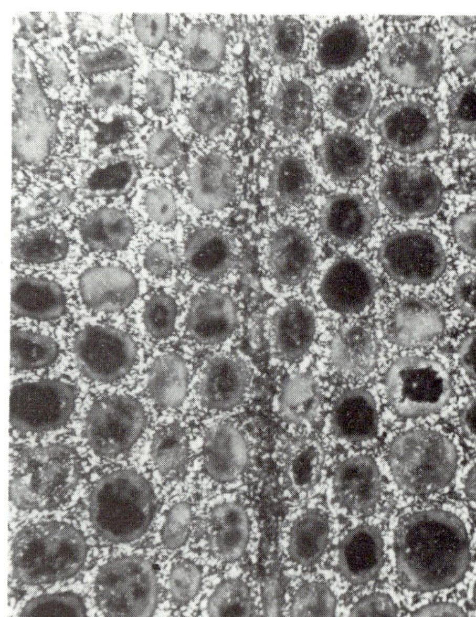

46b, LP 0.1 mm

Plate 10 *continued*

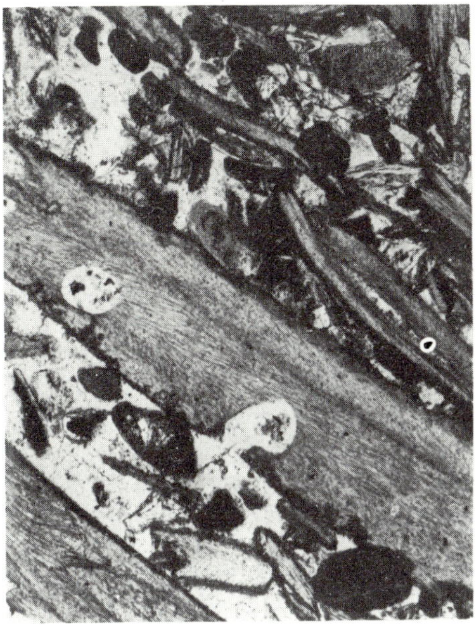

47a, LN 1 mm

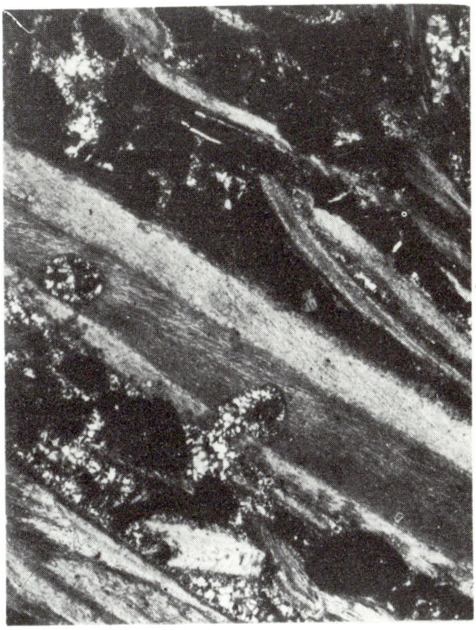

47b, LP 1 mm

45a and **45b**: In nodules from the Pyrenean Dinantian, it seems that the precipitation of microcrystalline apatite (clearly visible under polarized light) has been heavily influenced by radiolarian tests and organic particles.

46a and **46b**: Epigenetic alteration into microcrystalline apatite of a wood fragment of the Lower Albian at Maurupt-le-Montois, eastern Paris Basin (sample taken by P.J. Fauvel).

47a and **47b**: Detail of partial epigenetic alteration into phosphate of Ostrea coquina. The apatite formed at the expense of the carbonate is microcrystalline; its relatively high birefringence reveals a high level of PO_4 substitution by CO_3.

Togo. Other chambers or porous microenvironments may also play a role in the formation of phosphate particles. Tests of ammonites, pelecypods or gastropods have been known to provide moulds for at least some of the phosphate nodules of the Albian in the Boulogne region of northern France; among the gastropods, the pteropods provide internal moulds which have contributed in a big way to the formation of Eocene phospharenites (Togo, Guinea Bissau, etc.); algal structures have influenced the formation of Precambrian stromatolitic phosphorites in Rajasthan, India; they have also, according to Soudry and Champetier (1983), affected a considerable proportion of the Upper

Plate 11

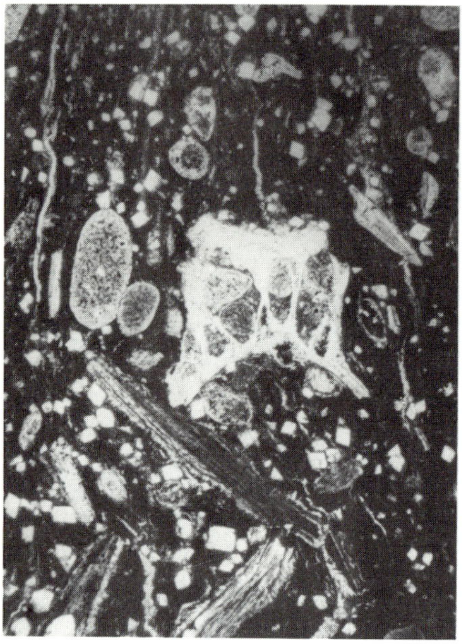

48a LN 1 mm

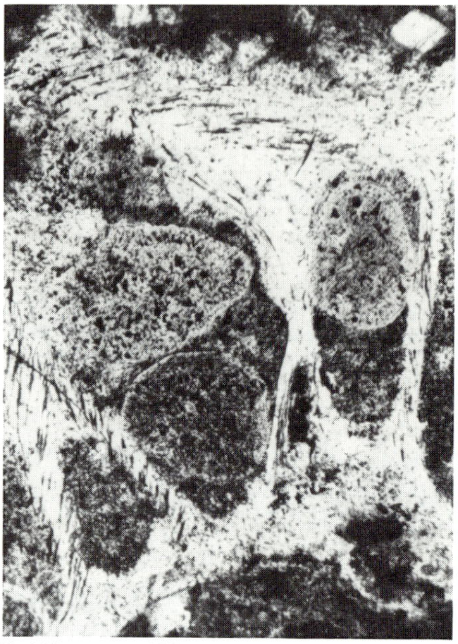

48b LN 100 μm

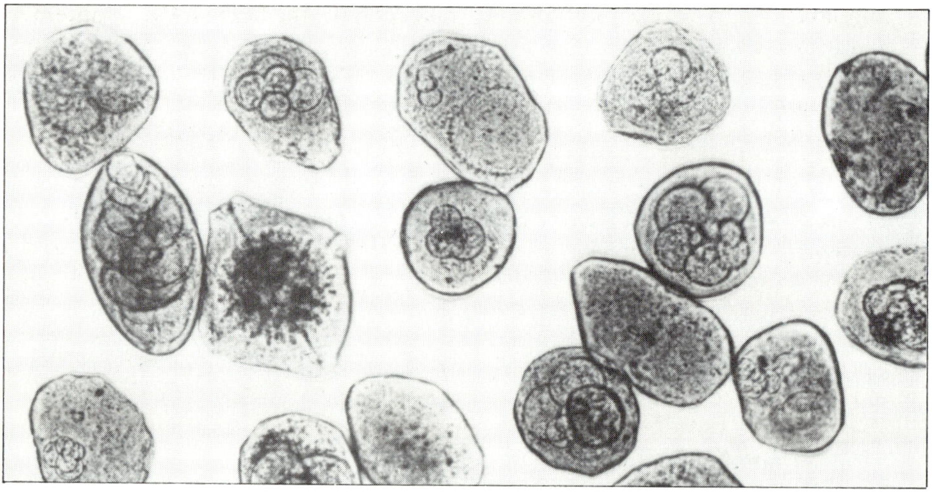

49 LN 100 μm

48a and **48b**: Tunisian Palaeocene. Pellets formed at the expense of the enclosing bioclast.

49: Middle Eocene, Togo. Pellets from the layer mined show by transparency after immersion in bromoform, the foraminifera and radiolaria they very often contain.

Campanian phospharenites in Israel. Phosphatization cells may be associated with these microenvironments. This was suggested by Lamboy (1976), following his investigation of the phosphorites of the continental shelf of north-eastern Spain. These 'phosphatization cells' are small porous 'containers' which are 'semi-confined' with respect to the general environment as a result of sedimentary or biological discontinuity where calcium phosphate crystallizes by direct precipitation or by epigenesis. Phosphate particles may develop in one or several contiguous 'phosphatization cells'.

The *role of faecal pellets*, i.e. coproliths, in phosphate sedimentation is not yet clear. There is general agreement that faecal pellets make a considerable contribution to the accumulation of organic matter on the sea floor, but there is still considerable debate as to whether or not they produce phosphate particles directly. Porter and Robbins (1981) attach a great deal of importance to faecal pellets of copepods in forming phosphate pellets of the Phosphoria Formation of the Permian in the Rocky Mountains; similarly, Champetier and Joussemet (1979a), and Lamboy (1982a) identify a great many coproliths among the phosphate grains of Lower Palaeocene–Eocene deposits in Tunisia. On the other hand, Cayeux (1950) felt that coproliths are present in phosphate deposits on a large scale only in certain specific cases, a view shared by the author.

According to Moore (1955), the faecal pellets of carnivores tend to show little consistency, while those of herbivores are firmer; the most solid are those of animals feeding on mud. Pellets of carnivores have the highest P_2O_5 content, which may reach 11% for sealions, according to Hutchinson (1950), but they generally disaggregate rapidly in water. Those of herbivores seem never to contain more than 0.2 – 0.4% P_2O_5, according to Moore, while those of mud-feeding animals only show high levels if there is already a considerable amount of phosphate present in the mud itself.

The sea floors where phosphate minerals are formed are not usually very favourable to benthic life and, therefore, to the abundant local production of faecal pellets. Mud-feeding animals do not appear likely to play any important role in the production of phosphate pellets except after the phosphate ha been deposited; this is what seems to have happened in Florida, where according to Riggs (1979b), as we pointed out earlier when discussing 'reworking', a benthic fauna including polychaetes and arthropods has contributed to the production of pellets from mud that was already largely phosphatic.

During the formation of phosphate minerals, the underlying sea floor seems to have been supplied with fragile pelagic faecal pellets, which require a very calm environment if they are to survive their phosphatization and induration. It is precisely these conditions that have been observed locally in the Eocene layer mined in Togo, where high concentrations of coproliths are associated with very well-preserved fish teeth, immediately above clays with dwarf microfauna and Tasmanaceae; this association suggests an 'accumulation trap' for dead fish and faecal pellets, under sufficiently calm and confined conditions for progressive phosphatization of the pellets at the expense of decomposition of the fish.

The *direct precipitation of phosphate* within diatomaceous ooze seems certain to be the way that the concretions described by Baturin (1970) on the southern African western continental margin were formed. Concretions develop at the expense of the surrounding interstitial water, whose P_2O_5 content falls below the saturation threshold. Burnett *et al.* (1982, 1983), in their work on the Peruvian and Chilean continental margin, describe nodules measuring several centimetres on the sea floor, whose lower surface is at present thickening by a few millimetres every 1000 years, drawing directly on phosphate in the interstitial water of the underlying mud.

Once started, therefore, it would appear that phosphate precipitation will tend to continue, thanks to a lateral flow of ions. The problem is one of deciding how the process starts. According to Burnett (1974), the process begins when the support surface onto which precipitation takes place acts as a catalyst to apatite precipitation. In the case of the Peruvian continental margin, this surface is a diatomaceous mud. Experiments carried out by Stumm and Leckie (1970) and by Kanel and Morse (1978), however, appear to indicate that the best surfaces are particles of calcite or aragonite. Observation appears to confirm that calcite frequently plays this role, since numerous phosphate pellets are formed around a Foraminifera nucleus whose test may avoid epigenesis. Other common surfaces include tests of Radiolaria, bone debris, quartz or feldspar fragments, dolomite crystals, etc. In a phosphate particle, apatite precipitation may be triggered by a single surface or 'nucleation site' or by a number of sites, particularly in the case of nodules (cf. Photograph 5 (Plate 1), and Plate 10). The cysts of dinoflagellates, which are often very common in deposits, are also probably good nucleation sites, but it has not so far been possible to identify them with certainty within grains.

What precisely is the role of the surface in apatite precipitation? In glauconite formation, Odin (1975) thinks that the porosity of the surface is essential to create the partial restriction necessary for ionic concentration. Similar reasoning may be applied to calcium phosphates, as shown by Lamboy (1982a), for whom the granular microenvironment represented by faecal pellets greatly assist the formation of phosphate grains.

However, the very considerable quantities of organic matter associated with such deposits, and the way in which it is often found inside phosphate grains, suggests that it too may play an active role in apatite precipitation. This idea is also confirmed by the internal bonds between humic acids and phosphates, discussed above.

The role of organic matter might be a consequence of the very high adsorption capacity of humic compounds towards various minerals such as quartz, calcite and the clay minerals (Hedges, 1980). The formation of a phosphate grain seems therefore to begin by the coating of an appropriate surface with humic acids, and the precipitation of the phosphate is then triggered by the acids themselves and not directly by the surface. This hypothesis provides a plausible explanation for the diversity of surfaces capable of producing phosphate particles; it may be extended to cases of apatite precipitation inside chambers, and suggests that phosphate pellets form around no more substantial a seed than a humic acid molecule, or a group of such molecules.

This hypothesis concerning the role of humic acids in the genesis of phosphate grains is made all the more plausible by experiments carried out by Suess and Futterer (1972) and Davies *et al.* (1978), who show that by adding humic acids to seawater enriched with carbonates, it is possible to produce generally spherical grains containing alternative layers of aragonite or magnesium calcite and humic organic matter, under laboratory conditions and without stirring; ooliths of this type have been produced both on quartz and calcite surfaces, and even without any mineral support. According to Davies *et al.*, these grains develop as a result of the alternating action of hydrophobic organic molecules of relatively low molecular weight, and organic membranes of high molecular weight. The hydrophobic molecules either form the initial seed or become attached to a mineral nucleus, and are then coated with the organic membranes, whose external surface contains sites favourable to the precipitation of calcium carbonate; growth of carbonate minerals is then guided, slowed down and eventually stopped by the hydrophobic molecules, which gradually become attached to them, and only resumes when new membranes have coated the entire system again. It should be noted that, according to Davies *et al.*, magnesium is indispensable to the formation of organic membranes, which in turn play an essential role in the build-up of successive layers.

As would be expected, the abundance of humic acids in natural phosphate particles has naturally led to attempts to manufacture phosphate grains in a similar way. Favourable results have been obtained from artificial seawater without magnesium, enriched in phosphate, carbonate, fluorine and humic acids extracted from Eocene phosphate rocks of the Gafsa Basin (Belayouni *et al.*, 1982). The apatitic grains formed, about 100 μm in diameter, only appear in the presence of humic acids, traces of which are also found inside the grains. The grains obtained are not concentric in structure, probably because of the absence of magnesium. In the presence of magnesium, i.e. in natural seawater, crystallization takes place more slowly. In experiments currently underway, it has been found that hydromonocalcite develops first, initially as twisted crystals, then in the form of rounded grains 50 – 100 μm in diameter, with concentric layers, after which the birefringence of these ooliths begins to decrease. Examination by scanning electron microscope shows that they contain calcium, magnesium and phosphorus in close combination. The apatite, however, is not immediately observable by X-ray diffraction. Experiments carried out by Gulbrandsen *et al.* (1984) suggest that it may develop only slowly.

Of the five types of mechanisms discussed, it is not easy to decide which is the most common, as phosphate particles in deposits have frequently been reworked since their formation, so their initial state is often not clear. What can be said, with some caution, is:

1. In the formation of phosphate particles, direct precipitation on a mineral or organic surface, or within a chamber, is a frequent occurrence. Organic matter, particularly marine humic compounds, play an important role in this precipitation. The supporting surface or surfaces may often be more or less completely epigenized by the apatite, but epigenesis is often not the only phenomenon involved in the formation of the grains.

2. Not enough is known yet about the role played in the formation of phosphate particles by living organisms, but it is clearly of some considerable importance. In particular, they influence the formation of these particles by providing a surface or a mould for apatite precipitation, or by supplying suitable microenvironments. Phosphatic coproliths derived from faecal pellets, on the other hand, do not seem to be a major source of phosphate, except in a few specific cases.

3. It is also only in certain cases that reworking is a major factor.

Finally, it should be noted that the more or less spherical, ovoid or rounded shapes of phosphate particles are generally natural and only rarely result from secondary abrasion in a high-energy environment.

3. FORMATION OF ORE DEPOSITS

The conditions under which phosphate particles may accumulate in sedimentary basins to form phosphatite deposits with economically viable grade and size may be discussed under the following headings: rate of phosphate sedimentation, influence of the low rate of phosphatite sedimentation on the facies, periods favourable to phosphate deposition, and regions favourable to phosphatite accumulation.

3.1. Rate of phosphate sedimentation

All phosphate ore deposits are characterized by a low rate of sedimentation. In Morocco, for example, phosphate sedimentation is mainly concentrated in the zone between the upper part of the Maastrichtian and the top of the

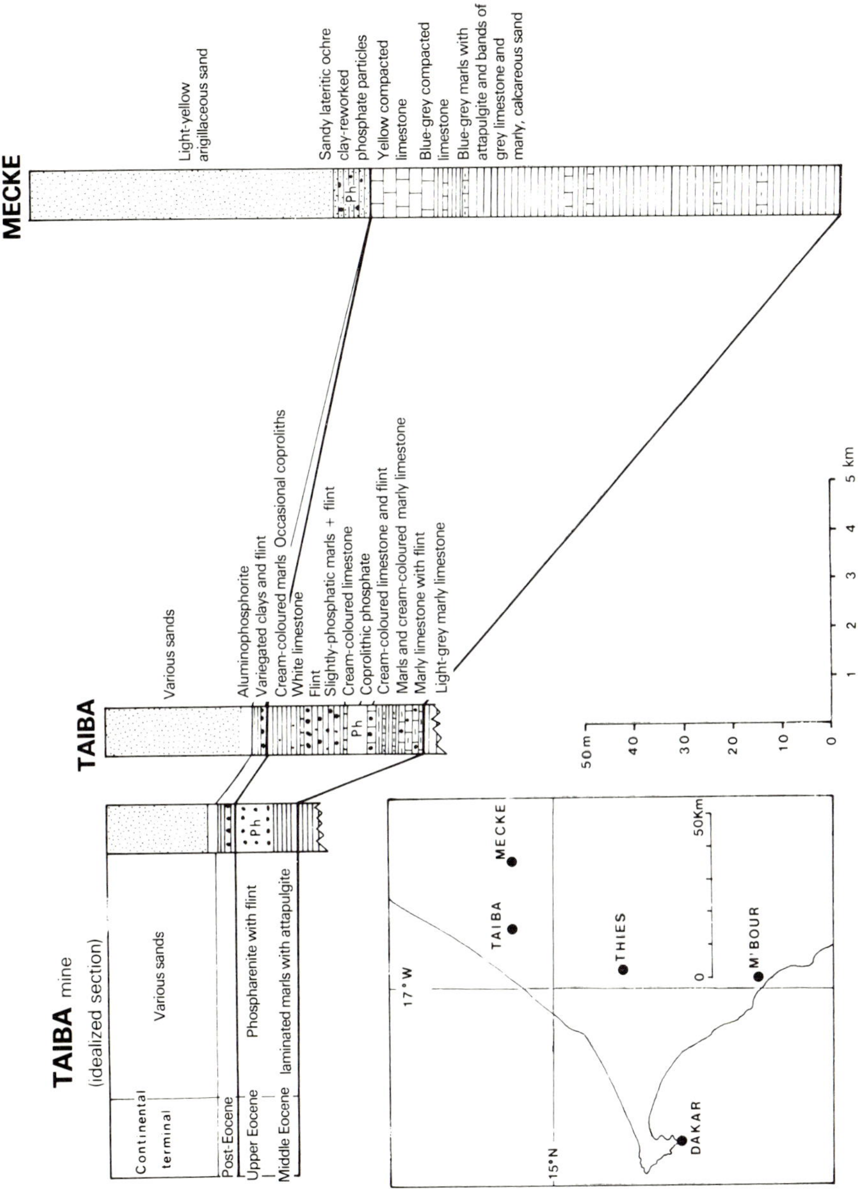

Fig. 20. *Reduction and disappearance of phosphate mineralization with increasing thickness of the Middle Eocene in the Taïba–Mecke region, Senegal* (from Monciardini, 1966).

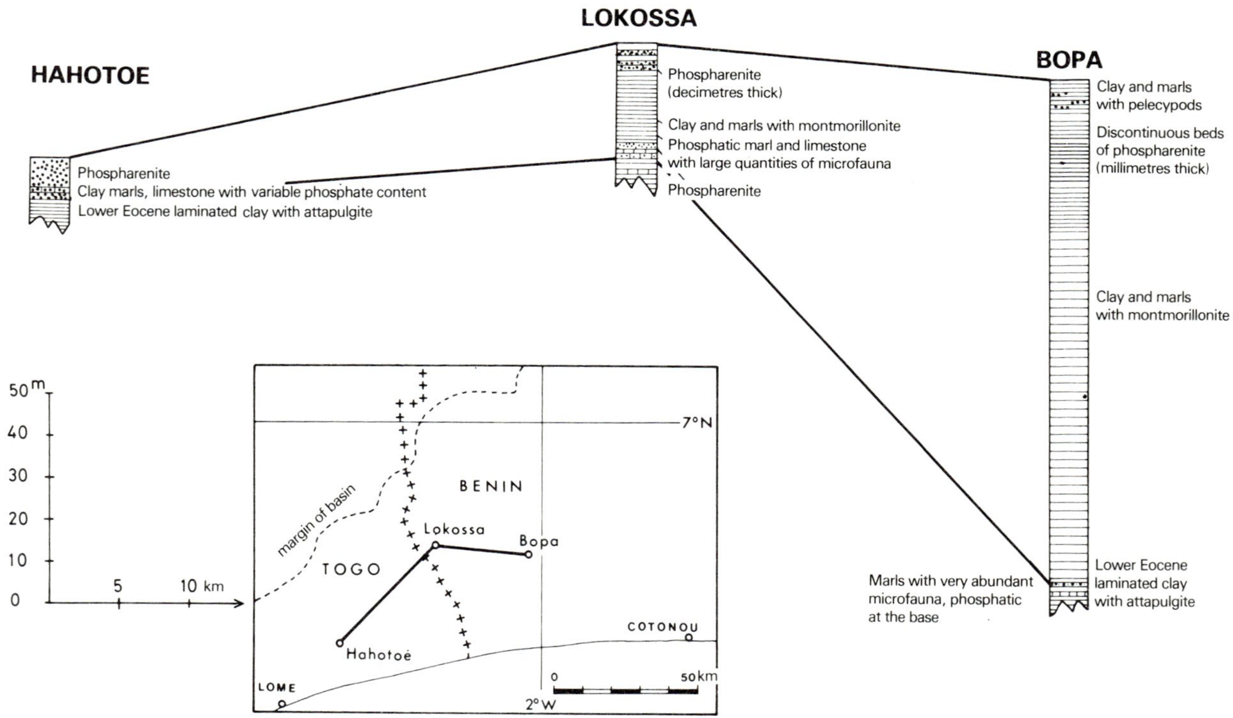

Fig. 21. *Development of the phosphatic Middle Eocene formation in Togo and Benin as a function of thickness.*

Ypresian, i.e. over a period of 15 – 20 million years. The thickness of this series varies (Boujo, 1976) from 25 m in the Khouribga region, which is the richest in Morocco, to about 80 m in the Youssoufia region. In Senegal, the Middle Eocene (about six million years) is about 12 m thick in the ore deposit currently worked; Fig. 20 (based on the data supplied by Monciardini, 1966) shows that the phosphatic fraction progressively disappears with increasing thickness of the series. In the Togo deposit, the phosphate sequence, about 10 m thick, represents an interval of time about ten million years long, straddling the Lower and Middle Eocene; as in Senegal, mineralization stops as soon as the thickness increases (Fig. 21). The same phenomenon is found in the Rocky Mountains, where the Permian B is about 30 m thick in its most highly phosphatic zone, although it can exceed 1000 m in regions without phosphate (McKee *et al.*, 1967).

These findings provide an 'order of magnitude' estimate for the rate of phosphate sedimentation: 200 000 to 1 million years per metre of deposit. This is an average rate and is not constant, as is made clear by sequential variations and the frequent discontinuity of phosphate formations. A more theoretical estimate by Baturin (1970), assuming phytoplankton productivity levels corresponding to those found today on the western continental margin of southern Africa, and optimized utilization of organic phosphorus, suggests a maximum phosphorite accumulation rate of 1–2 m per 100 000 years. Finally, on the Peruvian continental margin, the growth of phosphate nodules is estimated by Burnett *et al.* (1982) at a few millimetres per 1000 years (0.4 to 10 mm for 5 nodules), while the rate of accumulation of the neighbouring non-phosphatic deposits, which reaches about 1 mm per year, is several hundred times higher. The difference would be far greater if volumes were compared.

Despite the fact that these estimates are only approximate, they clearly reveal that phosphate sedimentation is a slow process. This has a number of consequences:

1. Periods and regions with rapid sedimentation, particularly with high levels of terrigenous supply or significant subsidence, are unsuitable for prolific phosphate sedimentation.

2. The greater the stability of the humic acids that control it, the more easily phosphate sedimentation takes place. As this stability depends on the nearness of the water–sediment interface and requires a moderately oxygenated environment, it is important that the burial of the organic deposit should proceed very slowly, and that the O_2 minimum layer should be close by throughout. In other words, the formation of a rich phosphate deposit is closely associated with condensed sedimentation and a high degree of stability in the palaeogeography of the basin and its supply of organic matter.

3.2. Influence of the low rate of phosphatite sedimentation on the facies

The slowness of phosphatite accumulation does not allow it to compete with terrigenous supply. The phosphate can therefore only accumulate where this detrital sedimentation is noticeably slow or absent, as a result of a dry climate or the presence of relatively flat land. The facies associated with rich phosphate beds may therefore include fine siliceous sands, silts or detrital clays with kaolinite, illite, etc., and in particular organically generated rocks produced by development of pelagic and (locally) benthic life.

Among rocks of pelagic origin, facies rich in silica or silicates generally predominate when the nutrient supply is abundant, as diatoms, dinoflagellates and frequently radiolaria are the predominant forms of plankton. Conditions are apparently more favourable for coccolithophorid proliferation if the phosphate supply is less abundant, and micrites with coccoliths seem to be more closely associated with relatively poor phosphate sequences, as in the Senonian chalk of the Paris Basin.

Silica of planktonic origin contributes to the production of cherts, and eventually to deposits containing clinoptilolite, and clays and marls with minerals generally rich in silica. Cherts have facies that range from blocks of different sizes found, for example, in the Taïba deposit in Senegal, to continuous beds, which may be isolated or contiguous, and may develop over a thickness of several metres or even several decametres if they are contiguous, as in the Colombian Senonian sequence. In the phosphate series of the Upper Cretaceous or Tertiary, the clay minerals formed under the impact of pelagic supplies mainly consist of smectites, attapulgite, sepiolite and green minerals fairly closely related to glauconite. Attapulgite and sepiolite, however, appear to be absent from the older phosphate series, the smectites being much rarer, while illite generally predominates. This mineralogical difference may be attributed both to the greater influence of detrital sediments and, according to Dunoyer de Segonzac (1969), to the effect of more pronounced diagenesis due to burial.

The benthic organogenic facies associated with the phosphate series are generally calcareous, with an extremely varied micro- and macrofauna. Salvan (1954), for instance, describes some 300 species of molluscs associated with the Moroccan ores. Among the molluscs, Ostrea sometimes predominate to the point of forming biostromes many decimetres or even several metres thick, as in Tunisia, the Nile Valley and Syria. Note too that advantage is taken in living systems of the effect of phosphates on oyster development: superphosphates are added to oyster beds to increase productivity.

In the Permian Phosphoria Formation of the Rocky Mountains, in the Gros Ventre region of Wyoming, sponges play an important role in the formation of cherts, occurring locally in the form of a reef complex with more or less cylindrical columns about 12 m in height and extending laterally over a considerable distance; this development is completely biogenic, showing a large number of spicules in thin sections.

3.3. Periods favourable to phosphate deposition

Phosphate deposits are widely distributed through the stratigraphic column, and deposits or concentrations of phosphate are known in practically all geological systems.

In the Precambrian to Infracambrian, Cook and McElhinny (1979) have found phosphate sediments dating from four periods. The oldest (1800 – 2000 million years) includes the deposits in Rum Jungle, northern Australia, and the Upper Peninsula, Michigan. The next period (1200 – 1600 million years) is represented by the deposits in the Udaipur region, Rajasthan, India, and the Siberian deposits of the Yenisei Mountains. The third period (700 – 800 million years) produced phosphate deposits in China, central Siberia and Brazil (Bambui). Finally, the Infracambrian (620 million years) produced considerable phosphate deposits in West Africa, in the region of the 'W' of the River Niger, which straddles the States of Niger, Burkina Faso (formerly Upper Volta) and Benin; phosphate strata of comparable age have also been discovered in the eastern part of Senegal.

The Cambrian is represented principally by the large deposits of Queensland (Georgina Basin) in Australia, Karatau in Siberia, Yunnan Province in China and Laokai in Vietnam.

The Ordovician is represented by the Tennessee deposits, whose economic value is due to extensive weathering, and the medium grade Obolus deposits in Estonia. Thick but low-grade beds of this kind, containing Lingula shells, are also known in northern Argentina (Zapla), and probably extend into Bolivia. Ordovician occurrences have also been found in central Siberia, in the U.S.A. (Iowa, Utah) and also in France (Brittany), particularly in the Crozon Peninsula.

No economically useful deposit of Silurian age has been discovered. In France, phosphate strata of this age have been discovered by Ters on the beach near Brétignolles in the Vendée (Deflandre and Ters, 1970).

The Devonian is another period for which few ore deposits are known. However, occurrences have been found in Siberia, in the U.S.A., in France (Brittany) and in Great Britain. The best Devonian phosphate formation appears to be in Iran, in the Elburz Mountains and in central Iran, but these have not yet been exploited.

The best known carboniferous phosphate occurrences of the Dinantian period are in the Pyrenees, where they occur in characteristic subspherical or ovoid nodules. This phosphate-bearing formation extends widely across Europe – it has been found in Germany and Poland. Phosphate nodules of comparable age are reported in the Adjjer (Algeria) and the Fezzan (Libya). Thin phosphate layers are locally intercalated in the Mississippi's limestone of Utah, U.S.A.

The Permian includes the rich deposits of the Rocky Mountains, of which the bulk straddles Wyoming, Utah and Idaho. The more modest deposits of the Mussoorie region in India, in the Himalayas to the north of Delhi, have also been attributed, by Parwardhan (1978), to the Permian period. Apart from these deposits, the Permian terrains generally do not appear to be phosphate-bearing to any great extent: occurrences have been reported from the northern Caucasus and in Siberia; in Vietnam, there used to be small-scale working of the weathered part of Permian phosphatic limestone.

The Triassic is also poor in phosphate. The best known deposits are in northern Alaska (Patton and Matzko, 1959) and Libya, about 100 km to the south of Tripoli, where two lenticular phosphate layers a few tenths of a metre thick were discovered during work on the 1/250 000 scale map NI 33-13 of Tarabulus. These occurrences are described in the booklet accompanying the map.

The Jurassic is fairly frequently phosphate-bearing, e.g. in France, where phosphate nodules appear at various levels in the Liassic and in the Middle Jurassic. Occurrences have been found in Greece, in north-eastern Siberia, in the Moscow Basin, in western Canada, in Argentina (Neuquen Basin), etc. Mineralization also frequently continues into the Lower Cretaceous, as in the case of the low-grade Soviet mines in the Moscow-Volga plain, or in small deposits in Mexico.

The Upper Cretaceous and the Eocene are rich in phosphate deposits: in the western and northern coastal basins of Africa, throughout the Middle East, in the U.S.S.R. to the south of the Urals (with a large low-grade deposit at Aktyubinsk), in Greece (Epirus), in the Paris Basin, in Brazil (the Olinda deposit, near Recife), in Venezuela and in Colombia.

Oligocene sequences, on the other hand, are practically barren, whereas the Miocene is again a source of rich deposits, as on the eastern coast of the U.S.A. in North Carolina, South Carolina and Florida (the Florida deposit is generally regarded as derived from Miocene phosphatic material reworked during the Pliocene), in California and the Mexican Baja California, in northern Venezuela and in the Sechura Desert in Peru. Miocene phosphate occurrences have also been reported in certain African coastal basins, particularly in Gabon and Senegal. Finally, certain phosphate deposits of the present continental margins (California, Blake Plateau, Chatham Rise, etc.) may be classified as Miocene.

The Pliocene includes the formation, by reworking, of the Florida deposits. Phosphate sedimentation seems to have continued on certain continental margins up to the present day, as was shown by Baturin and Burnett for Namibia and Peru. The formation of deposits developing from guanos at Christmas, Nauru and Makatea Islands (Burnett, 1980b) may be regarded as belonging to the Plio-Pleistocene, or even including the end of the Miocene period.

The considerable variation discovered in the degree of phosphate sedimentation marking different periods prompted Cook and McElhinny (1979) to estimate quantitatively the stratigraphic distribution of phosphate deposits. Their analysis, based on about 180 deposits of very varied character, appears in the left-hand side of Fig. 22. The curve obtained emphasizes several periods favourable to phosphate sedimentation, namely the Cambrian, Ordovician, Permian, Jurassic and Lower Cretaceous, Senonian-Eocene and Miocene. If, however, we are only concerned with economic or 'subeconomic' deposits (Fig. 23), we obtain a curve representing the estimated distribution of reserves (right-hand part of Fig. 22) which, though it has much in common with the overall curve, also has certain differences. These show that the existence of a deposit of economic size is not simply a function of the number of known phosphate occurrences associated with the period.

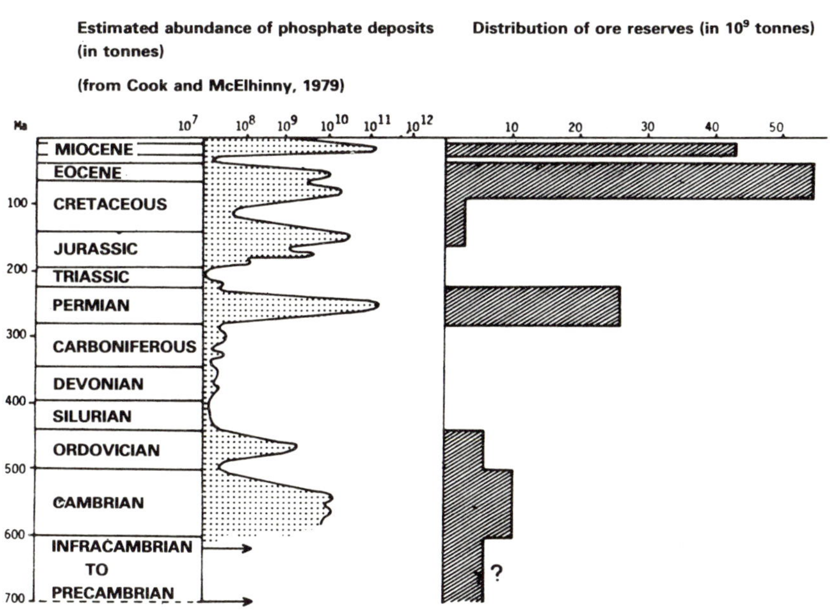

Fig. 22. *Stratigraphic distribution of phosphate deposits.*

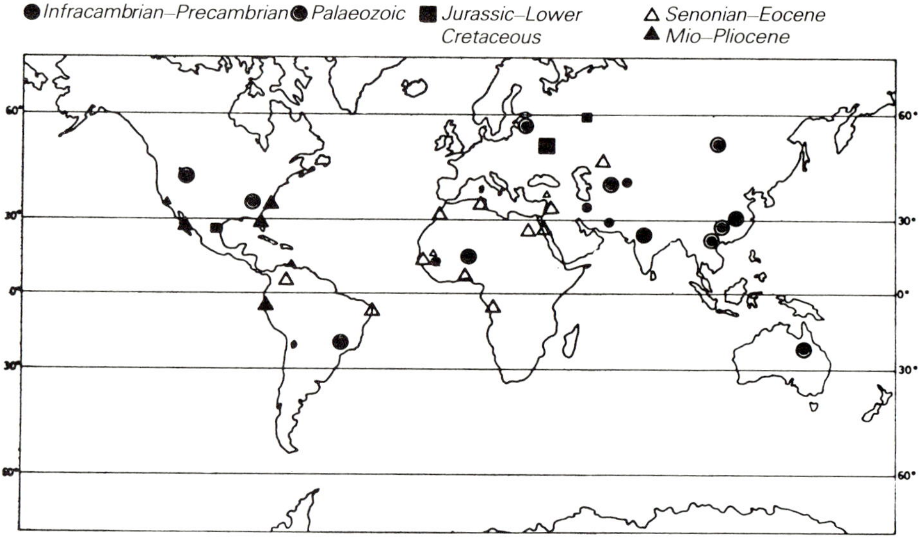

Fig. 23. *Location of major sedimentary phosphate deposits.*

Although, therefore, phosphate sedimentation is widely scattered throughout the stratigraphic columns, the known deposits provide clear evidence of the existence of privileged periods. What are the sedimentological and palaeogeographical characteristics of these periods? Work carried out by Fischer and Arthur (1977) provides some clues. These authors distinguish alternating 'polytaxic' and 'oligotaxic' periods in the history of the oceans. Polytaxic periods are marked by the presence of a wide variety of planktonic species, and by relatively uniform and high oceanic waters. Rising sea-levels let the waters extend over highly developed continental shelves abundantly supplied with organic pelagic sediments whose persistence is assisted by the considerable expansion of the O_2 minimum layer.

Oligotaxic periods show less variety of species, lower sea levels, colder waters, and higher temperature gradients favouring circulation within the oceans. More oxidizing conditions prevail on a reduced area of continental shelf. Conditions are less favourable to the conservation of organic sediments. The calcite compensation depth deepens. The present period is oligotaxic.

According to Fischer and Arthur, the polytaxic periods are the most favourable to phosphate formation, in particular because they cause increases in phosphate reserves in the sea, due to slower ocean currents, i.e. increased residence time of phosphorus. Figure 24 shows that these periods are marked by high palaeo-temperatures, and there is a good degree of correspondence between them and periods favourable to phosphate sedimentation in the Miocene, Eocene and Upper Cretaceous. The cooler temperatures of the

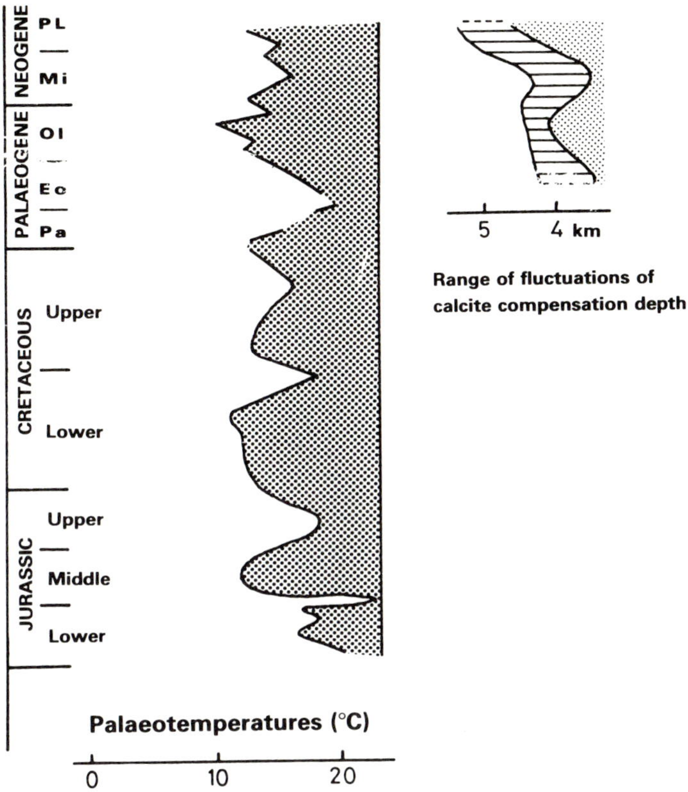

Fig. 24. *Stratigraphic variation in sea temperature and calcite compensation depth* (from Fischer and Arthur, 1977). *Palaeotemperatures were calculated using belemnites from north-western Europe up to the Maastrichtian; Globigerina from the South Atlantic were used for the Palaeocene, and from the tropical zone of the Pacific for the rest of the Tertiary.*

Oligocene also correspond fairly closely to a general shortage of phosphate. The correlation, however, is less satisfactory in the case of the Palaeocene, which produced major deposits in Morocco.

The (at least apparent) anomaly of the Palaeocene shows that the Fischer and Arthur hypothesis, which is also used to demonstrate that oligotaxic periods last some 32 million years, cannot be adopted without qualification as a way of identifying periods favourable to phosphate production. It does, however, cast light on certain aspects of phosphate sedimentation, when it takes place on a large scale.

The high level of the sea during polytaxic periods has a number of consequences:

1. It limits terrigenous supply in basins by the relative reduction in relief.

2. It increases the area of the continental shelf favourable to the formation of phosphate deposits.

3. It enables phosphate sedimentation to develop a long way from the deep sea regions, on the shelf or further in, as is the case of the Jordanian or Syrian deposits or phosphate occurrences in the Gao Basin, in Mali.

The increase in thickness of the O_2 minimum layer in the ocean, which may reach some 2000 m according to Fischer and Arthur, has the following consequences:

1. Development of laminated 'black shale' facies in the basin containing large quantities of organic matter of planktonic origin.

2. Storage, in these facies, of a considerable quantity of phosphorus which could be returned to the basin waters, still further increasing the nutrient contents of its upwelling currents. Lateral diffusion of the abundant phosphorus in the interstitial water may considerably accelerate the formation of phosphates near the pycnocline.

3. A considerable increase in bacterial denitrification processes, which, according to Piper and Codispoti (1975), will result in a reduction in the nitrogen/phosphorus ratio of the waters of the basin. This reduction influences the development of planktonic organisms, apparently leading to the spread of the phenomenon of dinoflagellates – rich 'red tides' (Pérès, 1976, p. 64) and the associated mass mortality of nekton, which in turn acts as a further powerful spur to phosphate sedimentation.

Finally, the high temperature of the sea is responsible for the abundance of benthic organisms, of the calcareous (Foraminifera, pelecypods, etc.) or siliceous type (sponges), on the sea floor above the pycnocline.

Phosphate formation is thus characterized by two main types of sedimentation: laminated facies rich in organic matter of planktonic origin, and organogenic benthic facies which are generally calcareous and sometimes siliceous. Phosphate deposits occupy an intermediate position between these two types of facies, and are more or less influenced by one or other depending on whether they form below or above the pycnocline; they may also be laterally displaced or disappear altogether, as a result of the movement of water or deformation of the basin. Terrigenous sediments, generally fine, may also mix with these various facies to a greater or lesser extent. In general, these sedimentological features of the periods favourable to phosphate formation are found in many deposits.

From a palaeoceanographic point of view, warm waters with a moderate temperature gradient and a thick O_2 minimum layer seem *a priori* comparatively unfavourable to the development of the strong upwelling currents generally necessary for phosphate sedimentation. It seems likely that such currents can only develop if the relative positions of landmasses and seas are highly favourable to the general oceanic circulation from which they originate. In the Atlantic, for example, upwelling currents are associated with the general movement of water between the poles and the equator, following the eastern edge of the ocean on their outward course, and the western edge on their return. It is probable that this gyratory movement could only begin when the opening process in the Atlantic had reached a certain stage, during the Upper Cretaceous, and phosphate formation could only develop in the coastal basins from this period onwards. On the other hand, an east–west marine current associated with a north–south separation of the continents seems likely to give rise to powerfully phosphatogenic currents at a very early stage in the opening process, according to Cook and McElhinny (1979).

Palaeoceanographic information of this type is undoubtedly valuable in defining periods favourable to phosphate sedimentation, but in the present state of our knowledge it is generally difficult to apply.

3.4. Regions favourable to phosphatite accumulation

Regionally, the accumulation of phosphatite depends on the volume of the phosphate supply, the existence of environments favourable to an abundant formation of phosphate minerals, and the conditions governing the accumulation and evolution of the organic deposits within which phosphate particles appear.

3.4.1. Abundance of phosphate supply

Since phosphate sedimentation is generally closely connected with upwelling currents, it has more chance of being abundant when these currents are stong.

Various factors may produce local upwelling currents (McKelvey, 1963; Bouysse, 1969). The most important seems to be wind, which by driving surface water away from the coast, causes deep water to rise and move in towards the shoreline. This mechanism is particularly active on the western continental margins in tropical or subtropical latitudes, in the trade wind zone, with very dry or even desert climates: California, Peru–Chile, South-West Africa, Morocco/Western Sahara, western Australia. Figure 25 shows that present or recent phosphate deposits occur mainly in such zones. Monsoons can also produce upwelling currents on certain coasts.

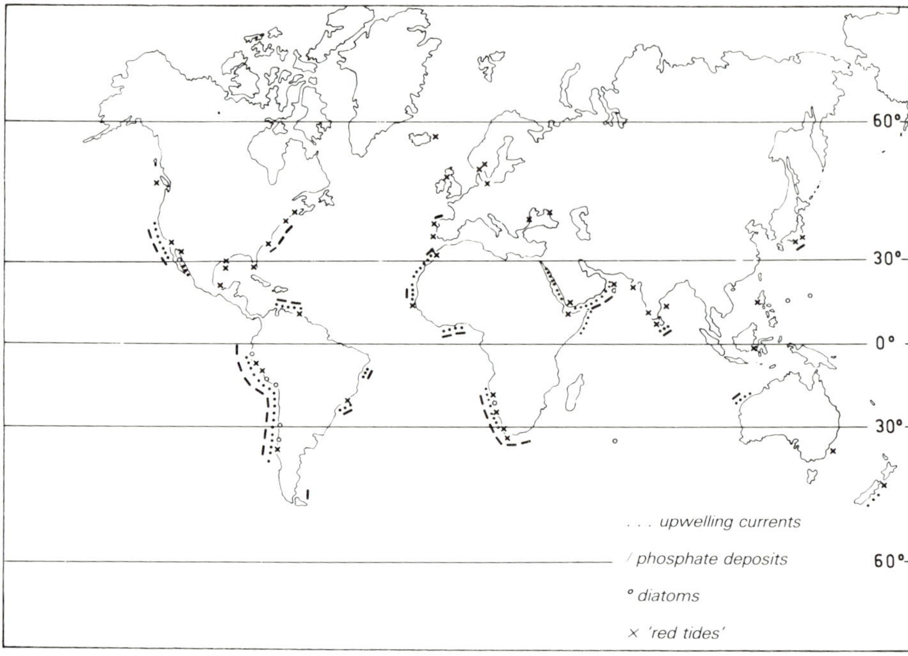

Fig. 25. *Location of present and recent phosphate deposits on the continental margins* (from McKelvey, 1967).

Upwelling currents may also occur on the eastern continental margins, if there is a warm current which draws surface water out to the open sea, if two currents diverge, at the boundary between two water zones which are different in nature, or, finally, if a seamount guides currents upwards from the bottom. Upwelling currents caused by seamounts seem to explain the significant phosphate occurrences discovered on seapeaks and guyots, as reported by various authors (Burnett, 1980a). Baturin *et al.* (1977), for example, describe phosphatic limestones of the Upper Cretaceous – Palaeogene, collected at depths of about 1000 – 2500 m in the north-west Pacific; they do, however, suggest that the water was shallower at the time of mineralization.

As McKelvey (1963) points out, upwelling currents which are the most effective in bringing about contemporaneous or Recent phosphate sedimentation generally occur at latitudes less than 40°. Sheldon (1964) and later Cook and McElhinny (1979) showed that the same applies to the older deposits, whose formation took place at low palaeolatitudes (Fig. 26). The palaeocontinental maps of Fig. 27, drawn from work by Briden *et al.* (1974), confirm this situation for deposits of the Precambrian, Permian and Senonian–Eocene; the map in Fig. 23 also shows that deposits of the Mio–Pliocene formed at latitudes less than 30° (the equator was at that time 5-10° north of its present position).

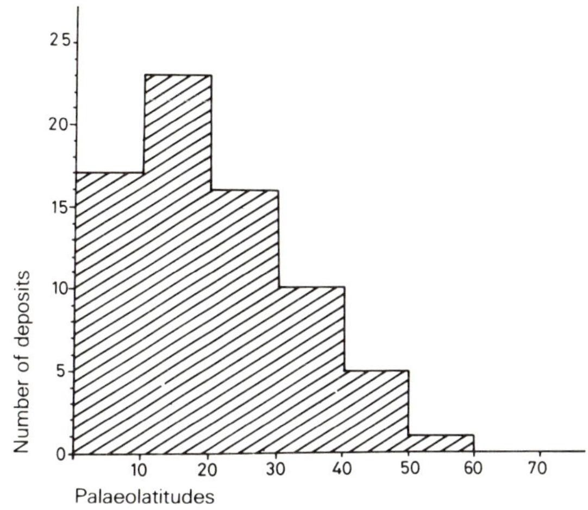

Fig. 26. *Palaeolatitudes of 72 major phosphate deposits* (from Cook and McElhinny, 1979).

Palaeolatitude is therefore an important factor in determining which regions were likely to show abundant phosphate supply.

On the other hand, when phosphorus is carried in larger quantities into a basin as a result of upwelling, plankton proliferates and, if basin depth is appropriate, produces abundant organic sedimentation, whose development leads, in turn, to the formation of localized zones of anoxic water and a reducing environment. In an environment of this kind, which is generally situated where the deposits are thickest, organic phosphorus will be gradually dissolved in the interstitial water where it remains in reserve as it cannot form phosphate minerals; it is then liable to travel quite widely in this state. In solution, it may travel vertically towards the basin, where it encourages further plankton proliferation, but it probably also travels laterally, which means that it may reach zones favourable of apatite formation and contribute to their enrichment in phosphate.

It is difficult to estimate with any precision the volume involved in this likely lateral supply of material to zones of apatite formation. It may be considerable, involving far larger quantities of phosphorus then are contributed by organic deposits in the phosphogenic environment. As sedimentation in this environment is locally very slow, it clearly receives less organic matter than do the adjacent zones of subsidence. In other words, an environment on which phosphate elements form will receive supplies of phosphorus in solution which will be all the greater, the more closely linked the environment is to thick anoxic organic deposits. This hypothesis is borne out by the palaeogeography of large world phosphate deposits such as those in the Permian of western U.S.A., the Eocene of Tunisia, Morocco, Senegal, Togo, etc.

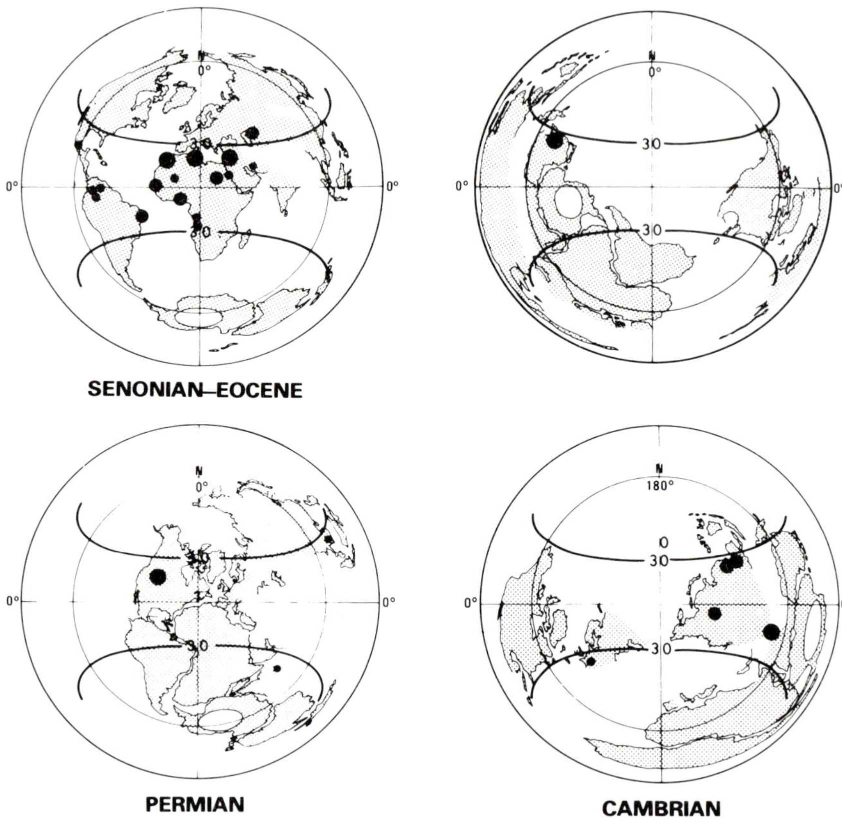

Fig. 27. *Position of major phosphate deposits on palaeocontinental maps of the geological periods in which they were formed.*

3.4.2. Location of phosphate deposits

The slowness with which phosphate accumulates means that an effective environment for phosphate formation will be in that part of the basin where the sedimentation rate is low, i.e. primarily at the basin's margin, or in or near a shoal area.

If it is to reach any size, the phosphate deposit will also require the presence of a 'trap' providing the kind of surface and the stable environment necessary for francolite to form from marine mud, and to protect phosphate particles from dispersion. This 'trap' may be: a local depression of the sea floor, due to erosion, tectonic movement or differential sediment compaction, a zone distant from currents or protected from them by a ridge on the sea floor, a part of the basin partially cut off by a submerged barrier, the interior of a narrow gulf, etc. The role played by the 'trap' is crucial in the formation of a deposit; it determines from the outset the limits to which it can extend. The relative wealth of phosphate associated with the different periods favourable to phosphate formation, as shown in Fig. 22, is undoubtedly related to a very great extent to the size of 'traps' that have enabled the organic matter to accumulate and to evolve as ore.

It has been shown above that the conditions favourable to the conversion of organic matter into apatitic minerals occur near the margins of the O_2 minimum layer in deep oceanic waters. In an 'accumulation trap' for organic matter, the oxygen content of the water generally increases from the bottom upwards, so we need only consider the upper limit. This is generally at between 50 and 500 m. In regions where plankton productivity is at present high, Burnett (1974), working on the Peruvian continental margin, placed the limit at about 100 m; the distribution of diatomaceous mud on the Namibian margin suggests that the limit here is comparable but shallower (60 m), and in the Santa Barbara Basin, off California, it may be as deep as 450 – 475 m (Sholkovitz, 1979, p. 39). Fischer and Arthur (1977) place it at about 150 – 200 m on their 'polytaxic' model.

An 'accumulation trap' is most effective for collecting large quantities of non-degraded plankton where the layer of water above is thinnest, so it is probably those areas where the upper limit of the O_2 minimum layer is nearest the surface that are the most favourable to phosphate formation. A similar suggestion is made by Bushinski (1964), who places the formation of all important deposits at a depth of between 30 and 200 m. The formation of deposits at depths of several hundred metres does, however, take place.

Even allowing for the possible contribution of lateral supply, a phosphogenic environment is unlikely, in general, to contain sufficient phosphorus concentrations to produce a rich phosphate stratum directly. Building up high concentrations of phosphate usually requires the deposit to be reworked after the phosphate particles have been formed, with the elimination of a large proportion of the organic matter, clays and various fine particles. This kind of winnowing, to which attention was drawn by Cook (1967) for the Ordovician deposits of Australia, may be associated with motion among the phosphate particles themselves, which therefore undergo a greater or lesser degree of fragmentation, wear and sorting. This natural enrichment of the deposit may be virtually *in situ* (e.g. by intertidal dynamic sorting) or may lead to displacement, with the phosphate particles being distributed according to the morphology of the sea floor. Riggs (1979b) suggests that the latter mechanism, which means that additional 'accumulation traps' become involved in the concentration of phosphate particles, though not in this case organic matter, explains the distribution of the Florida deposits: the microsphatite precipitated at the margins of the shoals by chemical or bacterial action hardens, disintegrates and is carried downwards in the form of intraclasts mixed with terrigenous material, along the sides of the shoals, where it accumulates in 'trap' basins. According to Riggs, the distance between the formation and accumulation sites may be as great as 100 – 200 km.

Such distances seem to be exceptionally high, however, as the palaeogeographic position of numerous deposits, on the margin of a basin or in the vicinity of shoals, suggests that phosphate particles have frequently accumulated near or actually at the site of the initial phosphate-forming organic deposit. Furthermore, Boujo (1976, 1980), who suggests that the distribution of the phospharenites of the Ganntour Basin in Morocco was governed by the nature of the sea floor, nevertheless feels that it is unlikely that the phosphate pellets were displaced significantly. The identical nature of the organic content of the barren and mineralized layers of the phosphate series of the Gafsa Basin also suggests that no such movement has taken place.

Baturin (1971) attributes the concentration of phosphate nodules on the Namibian continental shelf to periodical transgressions and regressions; during deep water periods, the nodules form in the mud on the sea floor, well away from the agitation of the surface waters; when the waters subside, the deposit will be affected by wave action, and the nodules are concentrated as the fine particles with which they were previously covered are washed away. A concentration mechanism of this kind seems suitable for various old deposits. It is also possible, however, that the necessary periodical agitation of seawater can be set up by the passage of marine currents over a barrier when the waters ascend or by any other factor causing the path of these currents to be modified.

4. SECONDARY EVOLUTION OF PHOSPHATE DEPOSITS

Once they have been exposed, phosphate beds are liable to surface alteration, the extent of which depends considerably not only on climatic conditions, but also on the petrography of the deposit and on its morphological and structural nature. The main transformations observed are oxidation of the organic matter, leaching of the exogangue, changes in the composition of the apatite and the formation of new phosphate minerals.

Oxidation of the organic matter depends not only on climate but also on the extent to which the organic matter itself has evolved. For example, the phospharenites of Morocco, Tunisia, Egypt, etc., which are grey or nearly black well below the surface, are very light in colour at the surface and to depths of as much as a few tens of metres. On the other hand, the older phosphate rocks (Dinantian of the Pyrenees, Permian of the Rocky Mountains) generally retain their dark colour at the outcrop, except when they have undergone surface weathering at an early stage, such as the Cambrian phosphatites of Queensland.

The gangue of phosphate beds outcropping on slopes is often more or less completely eliminated by leaching if it is calcareous or argillaceous. Phosphate particles concentrate and spread out, giving a false idea of the content, thickness and cohesion of the bed. Superficial study of natural outcrops may lead to a serious overestimate of the value of a phosphate deposit.

Leaching of the exogangue may transform a poor phosphate bed into workable ore. This is the case of deposits on the western slope of the Tennessee Valley, which contains ore with 18 – 24% P_2O_5, the residual product of the surface weathering of Ordovician limestone whose P_2O_5 content does not exceed 5 – 6%. These deposits in non-continuous pockets with karstic morphology are currently being mined; even within the last few years, their reserves have been estimated at 85 million tonnes.

Altschuler (1973), in his seminal article on the weathering of phosphate deposits, shows that the dissolution of a carbonate gangue is followed by partial dissolution of the phosphate in the upper part of the weathering profile, which then epigenizes calcareous rocks of the base or precipitates directly on contact with them. This phosphate migration observed by Altschuler in Tennessee, Florida and South Carolina, seems to have been an important factor in the formation of deposits at Riecito and Lizardo in northern Venezuela, which derive from Miocene phosphatic limestones and have a concretionary facies clearly suggesting secondary precipitation.

When leaching becomes sufficiently advanced, weathering will generally also affect the apatite itself, principally by reducing the extent to which PO_4 is replaced by CO_3 in the lattice. Observations made in present-day or old deposits tend to indicate, as McArthur (1978) stresses, that unweathered sedimentary apatite always consists of a carbonate fluorapatite with a high CO_3 content, whereas in weathered phosphate strata, the CO_3 content is always far lower. For example, the CO_3/PO_4 ratio is about 0.30 in the unweathered strata of the Gafsa Basin, but 0.05 – 0.06 in the Tennessee deposits and 0.06 – 0.07 at Bou Craa, Togo and Taiba, which have all undergone significant surface weathering. More dramatically still, this ratio falls almost to zero in the small Mae Ta deposit in Thailand, which has undergone intensive weathering.

This evolution of sedimentary carbonate fluorapatite under the effect of weathering has been studied in Morocco by the Strasbourg Institute of Geology (El Moüntassir, 1977; Lucas *et al.*, 1979c), which has shown that in the Sidi Daoui deposit the CO_3/PO_4 ratio decreases from 0.27 in the unmodified facies to 0.06 in those parts of the deposit that have undergone weathering, conventionally referred to as a 'podzolized' deposit.

These authors suggest that there is a reduction in the fluorine content which parallels the loss of carbonate caused by weathering. The same is found in other deposits, e.g. in Venezuela, where the ore may locally contain less than 1% fluorine. McArthur (1978) also observes a reduction in sodium, strontium and sulphate contents associated with weathering.

Fig. 28. *Generalized section of the Taïba workings, Senegal.*

From a practical point of view, these weathering effects are of interest in that they naturally lead to P_2O_5 enrichment of phosphate particles, making it easier to produce high-grade commercial phosphate rock. When weathering is more advanced, as in tropical climates, the practical effect is reversed because in the presence of clays, apatite is transformed into alumino-calcic phosphate minerals (millisite, crandallite) and then into aluminous minerals (augelite, wavellite); this kind of lateritic weathering runs in parallel with the transformation of all clay minerals into kaolinite.

The Taïba deposit provides a good example of weathering of this kind (Slansky *et al.*, 1965). The stratigraphical sequence in which this deposit occurs is shown in Fig. 28: the finely laminated clay of the footwall is of the attapulgite

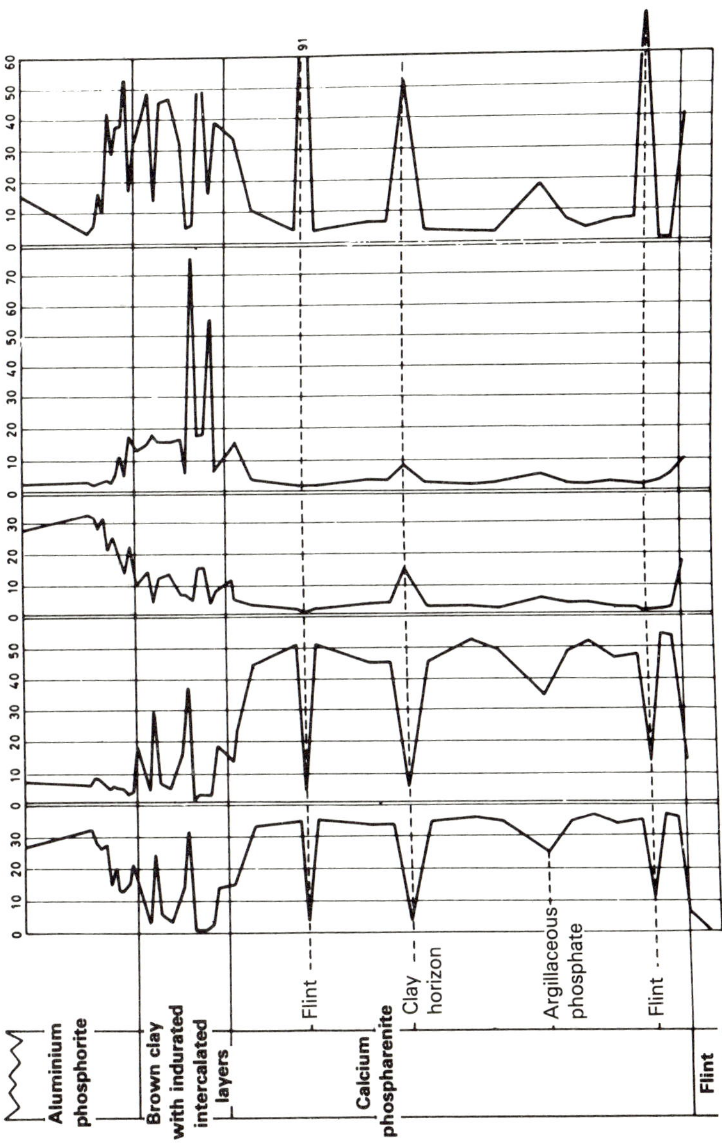

Fig. 29. Geochemical variation of the major elements in the aluminium phosphorites, the limit of brown clay and phospharenites at Taïba.

type, which gives way to montmorillonite just below the phospharenite ore; montmorillonite continues to predominate as far as the brown clay at the top. Further up, the argillaceous fraction evolves progressively from illite to kaolinite, while the phosphorite first becomes alumino-calcic (crandallite) and then aluminous (augelite). Figure 29 shows the main geochemical variations affecting the major elements of this section; at the level of the brown clay we find iron oxide enrichments, which may be associated with fine crystals or vivianite.

A detailed study of the aluminophosphorites of western Senegal by Flicoteaux (1982) distinguishes three main stages in their formation, starting from a parent rock comprising calcareous, argillaceous (attapulgite, smectite, illite) and phosphatic beds, with flints and sandstones.

There is a first, general, stage of development, under warm and damp weather conditions, dominated by leaching; it does not modify structures or volumes appreciably. Clay minerals become converted into kaolinite and goethite, while the apatitic phosphate minerals evolve progressively into millisite 1, characterized by the presence of aluminium, iron, calcium and sodium, into millisite 1 + crandallite 1, rich in strontium and barium, into cradallite 2 in which there has been less strontium and barium substitution and which has a 'sheaf-like' crystal structure, into crandallite 3, with lanceolated crystals, and finally into wavellite.

The second and third stages may not occur, and correspond to a climatic evolution towards increasingly long, dry periods. The second stage results from the introduction, in solution or in suspension, of silica, alumina, phosphate, iron and base elements; accumulations of kaolinite and of epigenetic products *in situ* result in an aluminous and then ferruginous duricrust. The third stage is the local chemical decomposition of the duricrust facies; the environment is relatively enriched in aluminium by reduction of its iron content; kaolinite and augelite form at the base of the profile, while calcite, apatite, chalcedony and quartz may also accumulate.

5. MECHANISMS OF URANIUM CONCENTRATION IN PHOSPHORITES

Concentrations of uranium in phosphate deposits may be present from the moment the phosphate particles form, or they may develop later. It is therefore important to distinguish between primary and secondary uranium concentrations.

5.1. Primary concentration

It is now generally accepted that seawater is the original source of uranium, in the form of dissolved U^{6+} linked to a carbonate complex. Uranium is integrated into apatite (Altschuler *et al.*, 1958; Baturin and Dubinchuk, 1974; Vceh *et al.*, 1974; Sheldon, 1959) by a process taking place in zones favourable to phosphate sedimentation: sediments on the sea floor are rich in organic matter, which can have the effect of concentrating the U^{6+} in seawater by several factors of ten, while reducing some of the hexavalent form to U^{4+} which can then be incorporated into carbonate fluorapatite during its formation; some of the remaining U^{6+} is adsorbed onto the surface of the apatite crystals.

In such cases, the amount of uranium coming into association with apatite will depend on various factors. The more plentiful the organic matter, for example, the greater the uranium content in the interstitial water and the greater the amount of U^{4+} that may be available for incorporation into the apatite lattice. Furthermore, as Sheldon points out, abundant organic matter will limit the dimensions of the apatite crystallites, which will therefore be more numerous, increasing the specific surface of the apatite and the possibilities for adsorption of U^{6+}. It is therefore likely that the uranium content of finer-grained phosphates will be higher than in the coarser type; this is confirmed by various deposits. For the same reason, bone debris that is porous tends to be far richer in uranium (several hundred ppm) than compact particles of the same size, such as coproliths or nodules.

The rate of sedimentation can also be a factor: low rates increase the time during which phosphate particles are in contact with interstitial water. The same effect may be produced if deposits on the sea floor are periodically set in motion by currents.

According to Gabshin *et al.* (1974), however, these mechanisms for the primary concentration of uranium do not seem to produce concentrations exceeding 300 – 500 ppm.

5.2. Secondary concentration

Soil development and water circulation may lead to the uranium being dissolved again. The effects of weathering of this kind depend on its intensity.

If there is considerable weathering, as in Florida or Senegal, apatite is dissolved in the upper part of the profile; some of the phosphorus escapes, while the rest combines with aluminium in clay to produce alumino-calcic or aluminous phosphates with which the dissolved uranium becomes associated once again. Observations at Taïba in 1961 showed that the alumino-calcic phosphates are richer in uranium than unweathered calcium phosphates, with a uranium content not exceeding 150 – 200 ppm; slight enrichment also occurs in the upper part of the unweathered calcium phosphate. In Florida, pebbles in the lower part of the profile, which become porous under the effect of weathering, may contain 560 ppm of uranium and sometimes as much as 2500 ppm, although their usual uranium content in this zone is 230 ppm (Altschuler *et al.*, 1958).

More moderate weathering does not transform the apatite mineralogically, and the phosphate that has passed into solution recrystallizes in the same form towards the base of the profile, where it becomes enriched in uranium. In the example from South Carolina cited by Altschuler *et al.*, the uranium content increases in this way from 5 to 740 ppm, and the U(ppm)/P_2O_5(%) ratio increases from 12 to 28. Part of the enriched zone, brought into prolonged contact with soil solutions under reducing conditions, even contained concentrations of uranium greater than 1000 ppm; the authors mentioned above believe that, in this case, it is likely that uranium was contributed by an outside source.

An external uranium supply does seem necessary if large quantities are concentrated. The phosphate deposit in such cases basically provides a trap for the uranium in solution in underground waters. If contact between dissolved uranium and phosphate lasts long enough, the uranium content of the underground waters may not have to be very high – the most important factor is the efficiency of the trap. For example, a structural depression through which surface or underground waters have drained over a long period into phosphate rich in organic matter and resting on a permeable bed, would provide an excellent trap.

6. EXAMPLES OF ENVIRONMENTS FAVOURABLE TO THE ACCUMULATION OF PHOSPHATITES

We shall now give some examples, drawn from the Upper Cretaceous–Tertiary of the African Atlantic coast and Tunisia, and from the American Permian Basin of the Rocky Mountains, to illustrate the points we have been making. It is not the intention to describe the phosphate deposits in detail, but simply to highlight certain characteristic features.

6.1. Coastal basin of Gabon

No really large phosphate accumulation has so far come to light in the coastal sedimentary basin of Gabon, but the basin is nevertheless of interest because of the palaeogeography of the local phosphorite concentration, discovered about twenty years ago by the B.R.G.M.

Phosphate occurrences have been discovered as a result of exploratory oil

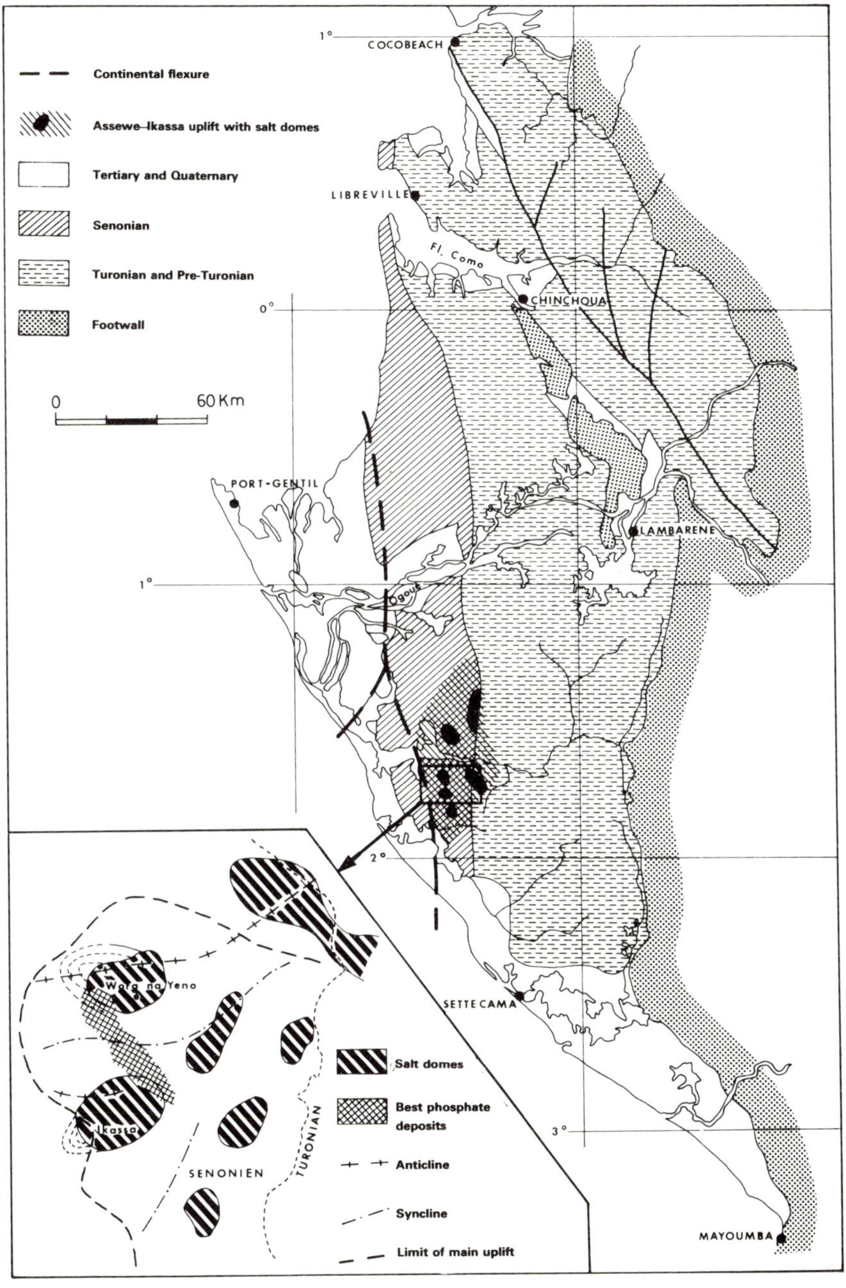

Fig. 30. *Location of Senonian phosphate deposits in the Gabon sedimentary coastal basin.*

drilling in the Lower Senonian, Palaeocene, Eocene, Oligocene and Miocene. It is only the Senonian occurrences, however, that are accessible at a shallow depth, and on which most of the prospecting has, accordingly, been carried out. The best phosphate concentrations have been found in the southern part of the basin, at the bottom of Coniacian clay and silt, and just above the Turonian calcareous dolomitic beds. Geological data obtained by oil prospecting have made it possible to locate these concentrations palaeogeographically on the Assewe-Ikassa uplift, which is on the shelf near the continental slope. The planktonic microfauna and ammonites which were collected near the uplift confirm that there was good communication between this zone and the open sea.

The uplift was produced by a number of deep salt domes that were active in the Coniacian, as shown in Fig. 30, and is based on an unpublished report by Maurice Donnot (1967). The best phosphate deposits appear to have formed and accumulated in the trap produced between two of these domes.

The Gabon basin therefore provides a good example of the role played by uplifts (the Coniacian series thickens northwards and the phosphate mineralization becomes diffuse) and by traps in the formation and accumulation of phosphatites. It also shows that halokinesis (i.e. salt movement) can contribute to phosphate sedimentation.

6.2. Benin – Togo coastal basin (Fig. 31)

The basin is the western extension of the Nigerian coastal basin. The Precambrian basement, which outcrops along its north-eastern boundary, has been traced by seismic measurement at about 1800 – 2000 m in the Cotonou region, and in the explored part of the basin (Slansky, 1962) the sedimentary series starts with the Maastrichtian, at first highly detrital and continental in origin but later marine with a clayey, marly facies. The Palaeocene becomes progressively less detrital, with a clay fraction predominantly composed of montmorillonite; the sequence generally concludes with highly fossiliferous limestone, often overlain by a phosphatic glauconite bed, at most a few tenths of a metre thick. This limestone suggests a fairly high energy environment. The Lower Eocene can be divided into three members, the outer ones being clayey–marly laminated shales with attapulgite, while montmorillonite predominates in the less finely bedded ore; the colour is black or dark grey at depth, owing to the presence of organic matter.

The Lower Eocene is fairly homogeneous where it is thickest. Elsewhere, both of the finely laminated members overlay a thin glauconite bed; other small calcareous glauconitic or phosphatic beds deposited in higher-energy environments produce interruptions at various levels in the upper members.

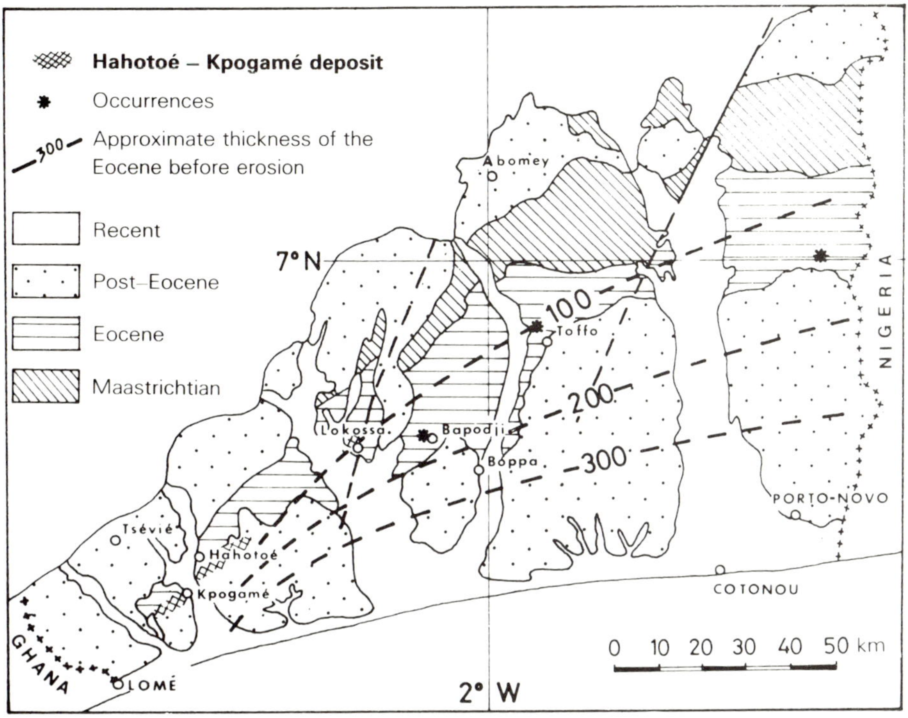

Fig. 31. *Generalized geology of the Togo–Benin coastal basin.*

The Middle Eocene is still fairly uniformly clayey – marly in the fully depressed part of the basin, with an argillaceous phase containing montmorillonite, progressively replaced by kaolinite towards the top; there are also fine sandy strata. Lower sequences contain marls rich in planktonic and benthic microfauna, which become more calcareous and phosphatic towards the margin of the basin as sedimentation becomes condensed. Figure 32 shows a section through an exploratory well which intersects this zone in the Bapodji region, Benin. The reduction in thickness in the Middle Eocene also leads to the appearance of phosphate pellets in the upper part, the quantities increasing towards the Hahotoe-Kpogame deposit, Togo, where they make up the bulk of phospharenite mined.

The Togo deposit is in a zone where the Eocene sequence is very thin, as is made clear by Fig. 21 and the isopachytes shown in Fig. 31.

In this zone the phosphate sequence rests on laminated clay shales (with attapulgite) formed in an anoxic environment. It begins with several metres of alternating marls and limestone, more or less phosphatic and often very rich in

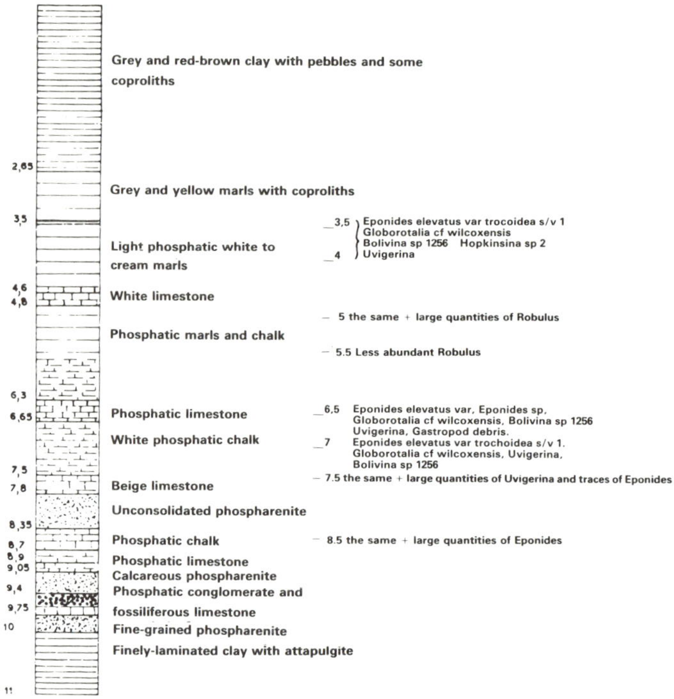

Fig. 32. *Phosphatic facies at the base of the Middle Eocene, in the Bapodji region, Benin* (from Slansky, 1962, p. 91).

Foraminifera. This lower part of the sequence generally also contains calcareous phospharenite beds, several centimetres or several tens of centimetres thick. A metre thick biophospharenite bed with internal moulds of pteropod shells may locally form the base of this sequence, deposited in an environment subject to very variable degrees of restriction.

These deposits, which vary considerably in their phosphate content, are overlain by a clayey–marly bed in which the wealth of organic matter with fossil algae (Tasmanaceae, Botryococcus), and locally of dwarf microfauna, indicates a return to a poorly oxygenated environment. This comparatively thin bed is constant throughout the zone and forms the footwall of the ore mined. The sequences above it comprise three main facies, distributed along an axis perpendicular to the margin of the basin.

The outermost facies comprises alternating beds of more or less calcitic phospharenites and phosphatic limestone or marls. This facies provides evidence of an unstable subsiding environment alternating between oxygen starvation and a varying degree of oxygenation; it is bounded to the north-west by a facies rich in biogenic limestone (Fig. 33) characteristic of a more stable and frequently well-aerated environment. This second facies, probably the result of a coastal current or a slight local upthurst, played an important role in the formation of the Togo deposit. Between this 'calcareous barrier' and the margin of the basin, as Kilinc and Cotillon (1977) point out, there is a confined trap which provides favourable conditions for the accumulation of a third facies, i.e. a phospharenite which is at present being mined. This depositional environment, which was comparatively stable, probably also constitutes the environment for the formation of phosphate particles, with the adjacent calcareous oxygenated uplift supplying the benthic foraminiferal tests very frequently found in the pellets.

The high quality of the ore mined by the Office Togolais des Phosphates (OTP) is also a result of surface weathering which has reduced the degree of carbonate substitution of fluorapatite in the pellets, while leading to the formation of only small quantities of aluminium phosphate minerals.

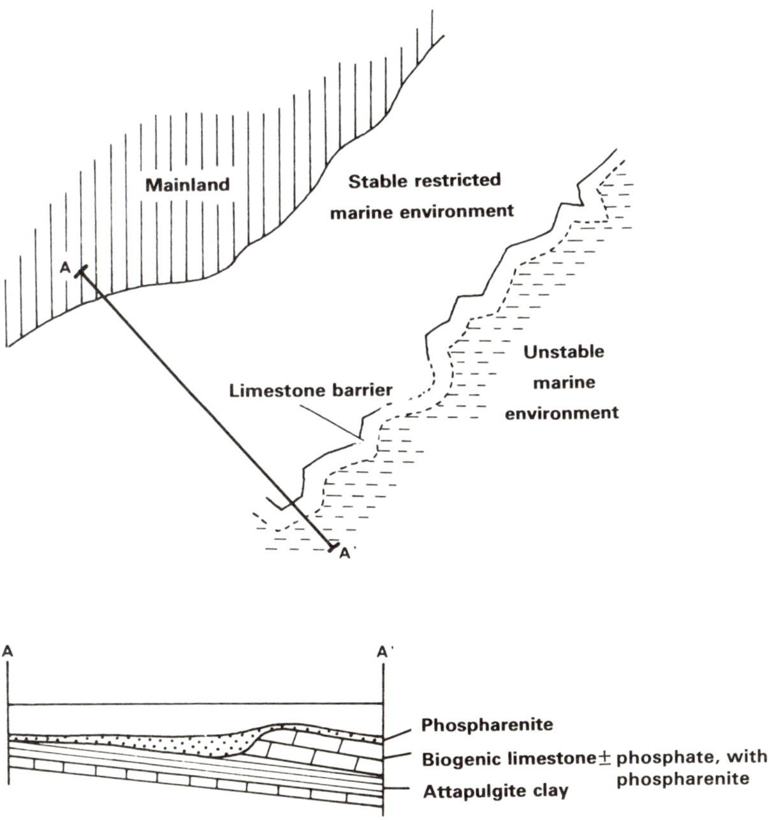

Fig. 33. *Depositional trap of Togo ore* (adapted from Kilinc and Cotillon, 1977).

It needs, finally, to be stressed that the phosphorites and the associated strata in the Togo–Benin basin, provide a good example of the influence of the environment on the nature of the deposit. The *in situ* marine formations of the Lower and Middle Eocene can be arranged, from the most reducing to the most oxidizing environments, as follows: clays and marls, firstly with attapulgites and then with smectites; phosphatites with coproliths and undisturbed fish teeth; pelphospharenites; pelphospharenites with calcite endogangue; glauconitic and phosphatic rocks; and, finally, biogenic limestones.

This classification is partly confirmed by Burnett (1980) from the continental margin off Peru and Chile, where nodules containing microsphatite, then phosphate pellets and finally phosphate and glauconite pellets, are distributed according to the concentrations of dissolved oxygen in the deeper waters, ranging from less than 0.2 ml/litre to 1 ml/litre. This distribution shows that, like the Togo coproliths, microsphatite nodules need a practically oxygen-free environment in which to form; it also confirms that the presence of glauconite in significant quantities indicates that the environment was slightly too rich in oxygen to allow large-scale apatite precipitation to take place.

6.3. Mauritania – Senegal – Gambia – Guinea Bissau basin

Phosphate occurrences in this very large basin have so far only been found on Senegalese territory and in nearby regions of Mauritania and Guinea Bissau. The stratigraphic section is essentially similar to the Benin –Togo coastal basin. The Maastrichtian, which starts with continental and detrital material, becomes progressively marine and less detrital, with more or less quartziferous limestones. The Palaeocene and Eocene are essentially 'chemical' in nature, with limestones, clays and marls containing montmorillonite, attapulgite and even sepiolite.

The Palaeocene and Lower, Middle and Upper Eocene contain a number of phosphate horizons. Small occurrences have also been found in the Miocene to the south-west of the basin. The Taïba deposit, a phosphate bed immediately overlying the marly attapulgite clay shales at the base of the Middle Eocene (Fig. 20), is mined about 100 km to the north-east of Dakar.

This bed is made up of phospharenite with intercalations of clay and chert, containing large flints which are more or less evenly distributed (Fig. 28); the top of the bed probably belongs to the Upper Eocene. A lens of phospharenite of the same age is also worked a little further to the south (Lam-Lam), as are aluminium phosphorites produced by the weathering of phosphatite beds.

Studies carried out at the B.R.G.M. by Monciardini (1966) make it possible to indicate the broad outlines of the basin's palaeogeography during the

Eocene. Its palaeogcography is marked in the Lower and Middle Eocene by two, partly emergent uplifts, the dome of N'Diass and that of Guiers Lake (Fig. 34), and two zones of subsidence in which planktonic microfauna indicate good communication with the open sea. The bulk of the phosphate occurrences of this period is distributed between zones around the uplifts and on the 'saddle-back' by which they are joined, and also along the shore of the Eocene sea, in regions which appear to have been in good communication with the waters of the great subsiding gulf to the south.

In addition to the Taïba deposit and the small deposit (of calcium phosphate) of Lam-Lam which are currently being mined, three further accumulations of potential economic value have recently been discovered in the basin, near Bofal in southern Mauritania, near the village of N'Diendouri to the south-east of Matam, in Senegal, and near Farim in Guinea Bissau. None of these ore deposits covers an area greater than 40 km^2, while the Eocene with abundant phosphate mineralization covers 30 000 km^2. This difference shows, on the one hand, that the mere presence of phosphate occurrences does not provide a sufficient basis for phosphate prospecting; on the other hand, it does indicate the importance of a trap in the formation of an ore deposit.

The trap that enabled the Taïba ore deposit to form is a triangular depression open towards the north-east (Fig. 35) and cutting into the north-north-eastern flank of the N'Diass dome. This 'trough' is probably of tectonic origin; its major axis, which runs mainly N50°E, is flanked by steeply dipping fractures.

The phosphate formation becomes thicker near the axis of the depression and near its opening, while at the same time it becomes enriched with calcite in the endogangue (with, in particular, incompletely altered Foraminiferal tests), and intercalated with calcareous beds. This lateral variation makes it difficult to determine the original nature of the ore now being worked, in which the CaO/P_2O_5 ratio is generally less than 1.45. Did it originally contain no calcite, or did it lose its calcite content by secondary weathering? It is certain that the ore has undergone weathering, since the carbonate substitution of its apatitic minerals is very limited, and it is capped by aluminium phosphorites; it is also clear, however, that the facies changes with increasing thickness (cf. Fig. 20). The question remains debatable.

The Farim (or Saliquinhe) deposit in Guinea Bissau has recently been studied in detail by a B.R.G.M. team lead by Prian (c.1987). This ore deposit is comparable in age with the Taïba deposit and, like the latter, is associated with a structural trap also running in a N50°E direction. The structure is shown in Fig. 36, and consists essentially of two palaeoridges with a depression about 4 km wide between them, interrupted towards the north-east and surrounded by uplifts on the other sides. The Farim structure has remained active since the formation of the deposit, locally producing displacement of strata.

The deposits at Bofal, described in detail by Boujo (1983), and at N'Diend-

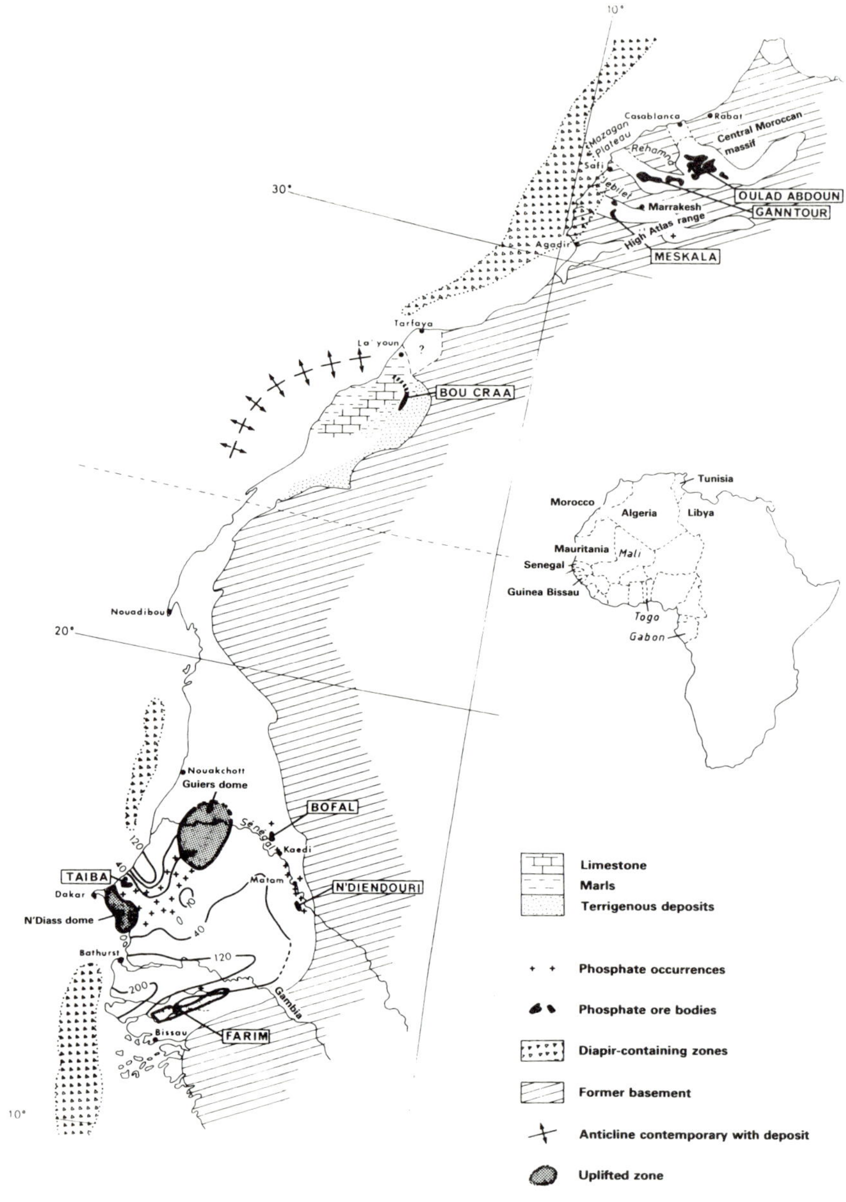

Fig. 34. *Palaeogeographic map of the coastal phosphatic basins in Morocco (Palaeocene–Lower Eocene) and Mauritania–Senegal–Guinea Bissau (Middle Eocene)* (from Von Rad and Wissmann, 1982; Boujo, 1976; Canto Romera, 1978; and various research projects by the B.R.G.M.). *In the great southern basin, isopachyte maps of the Middle Eocene, according to Monciardini (1966), and the uplifted areas provide evidence of the correlation between deposits and zones of restricted sedimentation.*

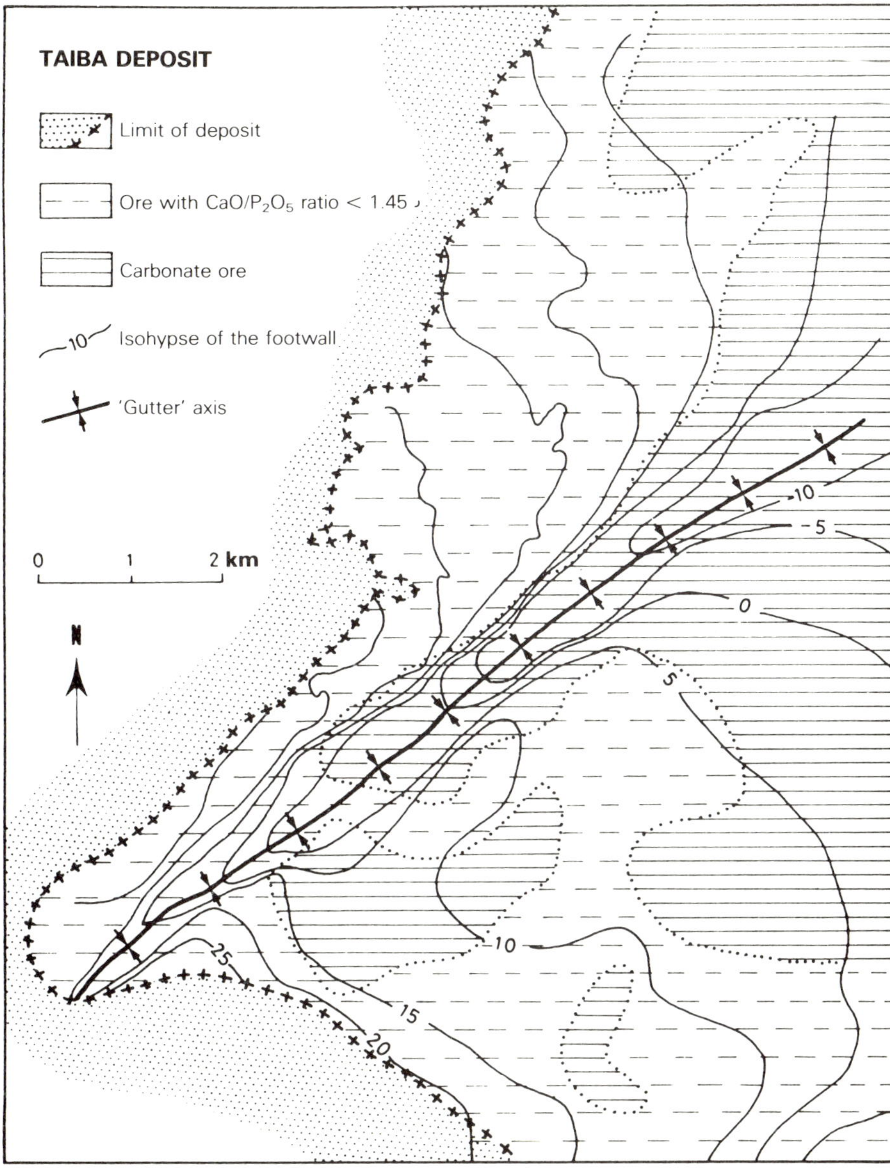

Fig. 35. *Structure of the Taïba deposit, Senegal* (from Atger, in Hameh, 1983).

ouri, studied by Pascal (c. 1987), are both immediately south of the terrigenous sandstones of the eastern margin of the Eocene basin. Their strata extend laterally westwards, to increasingly thick marly (smectite and attapulgite) and calcareous dolomitic sequences. The traps allowing the formation of these

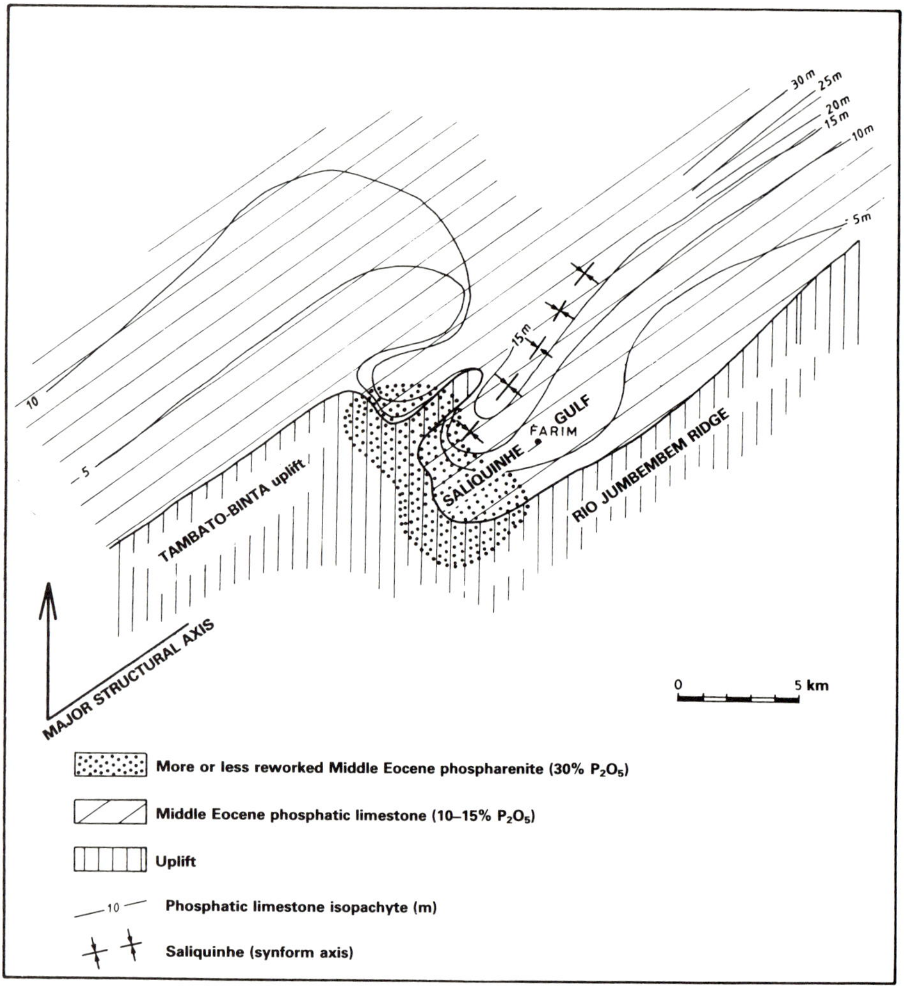

Fig. 36. *Structure of the Farim (Guinea Bissau) deposit* (from unpublished work by Prian, 1984).

phosphate ore deposits, whose age is comparable with those discussed earlier, have still not been precisely determined. It is possible that they are mainly hydrodynamic.

Finally, the recent oceanographic investigations, the results of which are summarized in *Geology of the Northwest African Continental Margin* (Hinz *et al.*, 1982), show that the continental margin of West Africa is marked with diapirs of Jurassic evaporites of which some at least have been active since the Mid-Cretaceous period. Zones with diapirs of this kind are to be found off the coasts of Mauritania, southern Senegal and Guinea Bissau, and they may have contributed to trapping the Eocene organic sediments from which the phosphate deposits of this period derived. The data available, however, cannot confirm this hypothesis.

6.4. Moroccan coastal basins

In the Moroccan coastal basins, phosphate deposits appear immediately above the Senonian marls, i.e. from the Maastrichtian onwards, and continue until the middle of the Middle Eocene; they are most extensive, however, in the Palaeocene and Lower Eocene. They are associated with more or less dolomitic or siliceous limestones, dolomites, cherts, and marls and clays with smectites/attapulgites. The phosphate series provides evidence of a marine environment very rich in organisms, including dinoflagellates, diatoms, Radiolaria, fish, etc.; it also reveals a very low rate of sedimentation (Salvan, 1959). The phosphate deposits are succeeded by a mainly calcareous sedimentation rich in gastropods and pelecypods, which occurred after the Middle Eocene during a general marine regression.

Morocco has more than half the world's resources of commercial phosphate, distributed between the deposit of Bou Craa of the La' Youn basin (Western Sahara), the Meskala region, where the phosphate beds distibuted among the basins of Chihaoua and Imi-n' Tanout are not yet expoited, the Ganntour region with its Youssoufia and Ben Guerir mining centres; and the Oulad-Abdoun region which has the biggest reserves, with its centres at Khouribga, Sidi Dahoui, Meraa-el-Ahrech and Sidi Hajjaj.

The immense size of the Moroccan deposits is certainly due to the intensity of marine upwelling. It is also due, however, to the size of the traps, which have enabled large quantities of organic material to accumulate and give rise to the development of major phosphate ore deposits.

In the basin of La'Youn (Ranke *et al.*, 1982; Von Rad and Wissmann, 1982; Cantoromera, 1978), which was apparently very open to the sea, the available data suggest that the depositional restriction required to enable organic matter to accumulate and for phosphate to form at the end of the Maastrichtian and during the Palaeocene periods has been produced by a number of factors (Fig. 34):

1. The gradual breakdown of communication, to the north, with the Tarfaya Basin.

2. An anticlinal ridge on the continental slope.

3. The offshore formation of limestone deposits that produced a restricted

> environment favourable to phosphate sedimentation in a depression some tens of kilometres wide between them and the terrigenous margin of the basin. In this phosphogenous environment, apparently situated, as in Togo, behind a limestone barrier, Canto Romera (1978) has shown a clear inverse relation between the P_2O_5 content of the phosphate formation and its thickness. The Bou Craa deposit is thus located on the eastern flank of this small, internal, confined basin.

The phospharenite mined is associated with beds of chert and clay; towards the north and north-west the sequence becomes enriched in limestone and increases in thickness; towards the south the ore becomes thinner, giving way laterally to siliceous sediments spread along the shore.

The Meskala, Ganntour and Oulad-Abdoun deposits occur in two narrow gulfs penetrating several hundred kilometres into the hinterland (Boujo, 1976, 1980). These gulfs are delimited by the High Atlas chain, which began to form as early as the end of the Cretaceous period (Stets and Wurster, 1982), and by the Palaeozoic massifs of Jebilet, Rehamna and Central Morocco. Depending on how the evidence is interpreted, it would appear that the Oulad-Abdoun gulf opened to the sea either immediately to the north-west or via the Youssoufia region.

Diapirs of Triassic to Liassic salt (Fig. 34), active since the beginning of the Cretaceous (Hinz *et al.*, 1982), have probably contributed to the large-scale accumulation of organic sediments on the continental margin and in the gulfs. A limestone plateau (the Mazagan Plateau) may have played a similar role up to the Palaeocene period.

Finally, general and local deformation of the floors of the gulfs, trapping organic matter, have also greatly influenced the formation and distribution of phosphate deposits (Boujo, 1980). The richest strata, for example, are Maastrichtian and Montian at Youssoufia, and Thanetian and mainly Ypresian at Khouribga.

6.5. Tertiary basin of Tunisia

In Tunisia, phosphate sedimentation took place primarily in the Palaeocene and Lower Eocene. During this period, marine sedimentation was influenced by a palaeogeography characterized by an arid climate; the Saharan platform emerged and the 'Kasserine Island' was penetrated to the north and east by a series of short gulfs and by various uplifts and small islands. The internal Gafsa Basin communicated to the west with the evaporitic Saharan basin, and was separated from the open sea by two straits which may have acted as bars. The palaeogeographic details are still open to various interpretations, particularly as regards the exact shape of Kasserine Island, its continuity with the Saharan platform and the existence of certain shallows. Figure 37 is principally based on the interpretation given by Fournier (1980), although we have added a strait to the south of Kasserine Island, according to Sassi (1974, 1980), who also identifies an uplift connecting the island to the region of Tunis. Burollet and Oudin (1980), moreover, mention the absence of the Lower Eocene to the north-east of the island, in two elongated zones shown by dotted lines.

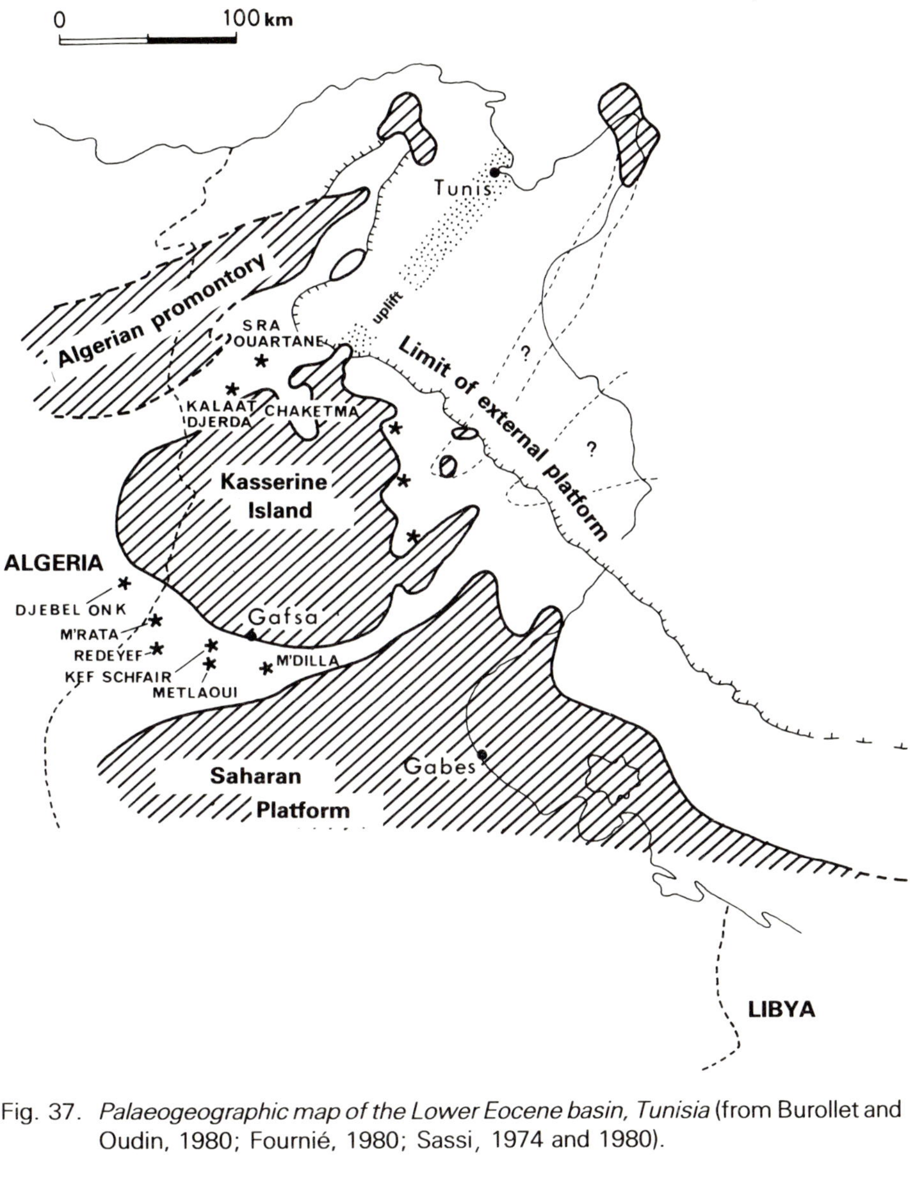

Fig. 37. *Palaeogeographic map of the Lower Eocene basin, Tunisia* (from Burollet and Oudin, 1980; Fournié, 1980; Sassi, 1974 and 1980).

According to Sassi, the sedimentation on the two sides of Kasserine Island from Palaeocene times differs. To the south of the island, the Palaeocene is first represented by slightly gypseous clays and marls of the El Haria formation and contains between one and three thin phosphate beds. The formation is 30 – 100 m thick. Montmorillonite is the main constituent of the clay fraction, and becomes associated with a little kaolinite towards the base.

The Metlaoui formation dates from the Palaeocene at the base and from the Lower Lutetian at the top. It contains three members:

1. The lowest member in the Oued Selja to the east of Metlaoui, is about 80 m thick. It is mainly dolomitic and gypseous, with beds of marls and shelly limestones. Montmorillonite is the main component of the argillaceous fraction.

2. The middle member corresponds to the main phosphate unit; it is about 40 m thick in this region and may contain up to nine phosphatic strata associated with clays, marls, shelly limestones (generally with Ostrea) and cherty facies. Laminated shale facies are frequent. Montmorillonite is still predominant but is frequently associated with attapulgite and sepiolite. Clinoptilolite is frequent. The organic matter associated with these facies is of planktonic origin; cysts of dinoflagellates are often very abundant, and cherty facies appear to have formed from a rich diatomaceous phytoplankton.

3. The uppermost member (Lower Lutetian) is about 60 m thick in the Oued Selja; it is mainly calcareous, with Ostrea coquina, the middle part being more marly and possibly phosphatic. Montmorillonite still predominates in the argillaceous fraction, together with attapulgite and illite.

The Souar formation (Lutetian to Priabonian) marks the end of marine sedimentation to the south of Kasserine Island. It is mainly gypseous and often dolomitic.

To the north of the island, sedimentation is less confined. At Kalaat Djerda, for example, the Souar formation is marly and the middle member of the Metlaoui formation rests directly on the El Haria formation. The upper member of the Metlaoui formation is often nummulitic. Although montmorillonite predominates in the argillaceous fraction, it is associated with glauconite, which may be present in large quantities (Chaketma, Sra Ouartane), rather than with attapulgite or sepiolite. There is no clinoptilolite.

The palaeogeographic diagram in Fig. 37 shows that the main accumulations are distributed around Kasserine Island: to the north, between the 'Algerian promontory', Kasserine Island and the uplift that is its extension towards Tunis; to the east, in small gulfs cutting into the island; and to the south, between the island and the Saharan platform. In the latter region (Gafsa Basin), communication with the sea is less direct and liable to bc interrupted, which explains why thc facies are often confined, and why evaporite occurrences are frequently found. To the north, on the other hand, communication is far better, the environment is less confined and unfavourable to attapulgite and evaporites, but favourable to glauconite, which develops preferentially to apatite.

Sassi (1980) has also shown that the main phosphatic unit of the Gafsa Basin can be subdivided into four successive lithological sequences. In each of them, the main phosphatic strata follow predominantly carbonate strata often containing shelly limestones, and are followed by clayey–marly or occasionally cherty beds. These sequences therefore seem to suggest that mineralization took place between periods in which the depositional environment was at first highly and then only very slightly oxygenated. When examined in detail, however, the sequence of facies is always more complex. Sassi has also shown that the overall trend is for the phosphate content to increase with the condensation of the series towards Kasserine Island, and that locally the distribution of phospharenites is controlled by the palaeoreliefs and deformation of the basin floor.

Thanks, therefore, to the detailed studies that have been carried out, the Lower Palaeocene–Eocene of Tunisia enables us to define a precise model for a basin with phosphate sedimentation. This model emphasizes, in particular, the links that exist between the abundance of phosphate mineralization and the absence or scarcity of terrigenous supply, the proliferation of phytoplankton associated with adequate penetration of nutrient-rich currents coming from the open sea, the existence of hydrodynamic traps enabling organic matter to accumulate, and a sufficient degree of restriction in the depositional environment. In these traps, the facies vary as the degree of restriction changes, and evaporites may appear under conditions of very severe restriction. Glauconite, on the other hand, develops where there is very direct communication with the open sea.

Finally, the nature and size of the trap have an important role to play: the palaeogeography of the Gafsa Basin, for example, is directly responsible for the accumulation of considerable reserves of phosphate ore, extending from the M'Dilla deposits in the east to the Djebel Onk deposit, Algeria, in the west.

6.6. The Permian of the Rocky Mountains, U.S.A.

A precise description of the Permian Phosphoria Formation in western U.S.A. has been given by McKelvey *et al.* (1953, 1956), Sheldon *et al.* (1967), Maughan (1975, 1980), etc. and there is no point in repeating the details here, but it is worth briefly mentioning the broad outlines of the palaeogeography that has given rise to such a large accumulation of phosphate.

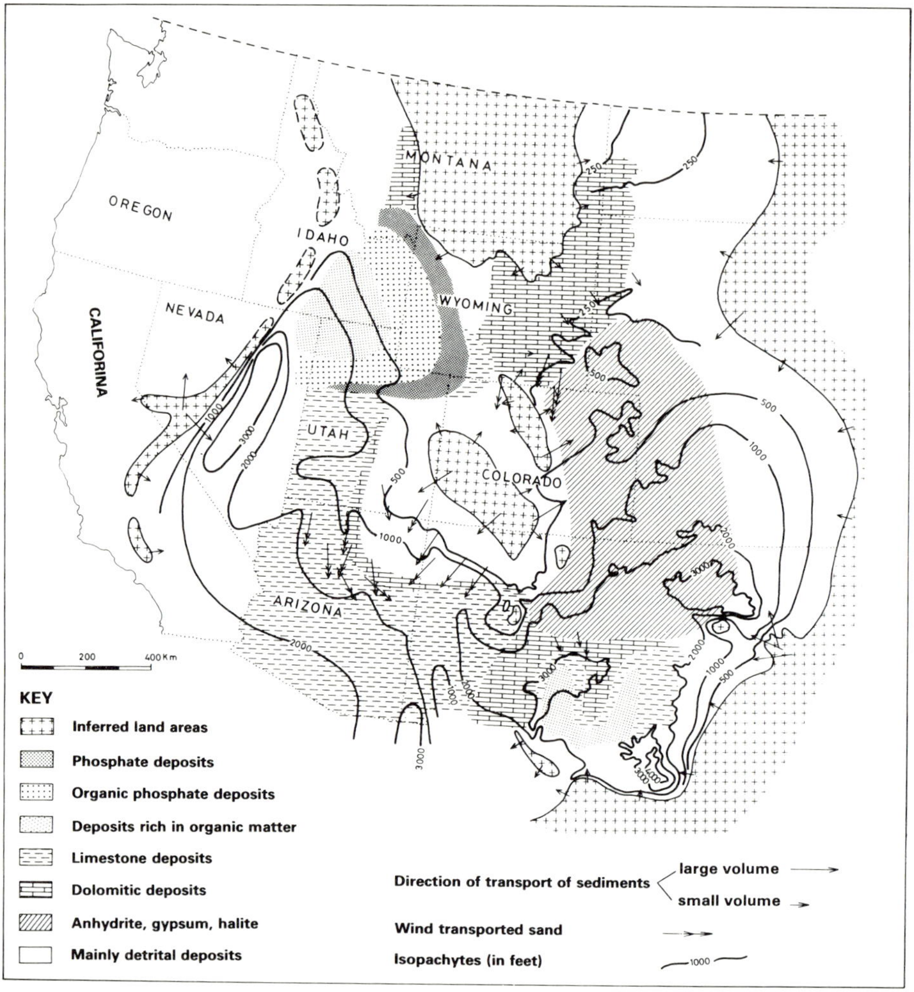

Fig. 38. *Palaeogeographic map of the western United States during the Permian B interval* (from McKee *et al.*, 1967). *Distribution of significant chemical sediments.*

The palaeogeographic diagram in Fig. 38, based on the work of McKee *et al.* (1967), shows the thickness variations of the Permian B interval of the Rocky Mountains equivalent to the Leonard Formation, which contains the main phosphate deposits. Lithological information is shown in the diagram by the distribution of the most significant chemical, and therefore also biogenic, fraction; the detrital terrigenous supplies, which are generally fine, are shown by arrows. These supplies locally influence the facies; a map of the lithofacies by Sheldon *et al.* (1967) shows, for example, that the most highly phosphatic crescent-shaped zone is bounded to the north-east by a parallel band of phosphatic sandstone and sandstone. Within the basin as a whole, however, detrital supplies seem to have played only a limited role during this arid period, as is made clear by the evaporitic sediments that develop to the east.

The phosphatites are associated with argillaceous and cherty facies, the entire sequence being rich in organic matter with an organic carbon content of up to 9% (Maughan, 1975). According to Sheldon (1966), this organic material is, moreover, the source of Permian oil extracted on a commercial scale further east. The phosphate beds are also frequently rich in tests of inarticulate brachiopods, and the cherts, which are rich in sponge spicules, are undoubtedly biogenic. In a personal communication, Wardlaw (1979) reports that these facies contain numerous Acritarchs, planktonic organisms comparable with dinoflagellates of the most recent phosphate deposits.

The isopachyte lines show that the phosphate deposits are on the continental platform, in a zone of reduced sedimentation. The basin as a whole is partly isolated from the open sea by a line of uplifts of various depths, creating a large hydrodynamic trap behind it, i.e. in the direction of the landmass. Nutrient-rich upwelling currents apparently from the north-east, and the relative restriction due to the trap, have formed an environment which is particularly favourable to the large-scale accumulation of organic deposits. These deposits could develop into phosphates wherever they are likely to undergo the necessary moderate oxidation and periodical agitation, i.e. in the shallower water zones.

The Permian of the Rocky Mountains therefore illustrates the close relationship between the nature of the trap and the extent of phosphate ores, a relation that has already been stressed with respect to the Tunisian and Moroccan Tertiary sequences.

6.7. Conclusion

By way of conclusion, the most important characteristics of phosphate sedimentation, shown in Fig. 39, may be summarized as follows:

1. Phosphate deposits are mainly supplied with phosphorus from deep oceanic reserves by means of upwelling currents. This supply of phosphorus presupposes satisfactory communication between the depositional environment and the open sea.

2. The phosphorus carried by upwelling currents stimulates the development of large quantities of plankton, which enrich the sediments with organic matter in the O_2 minimum layer of the basin. The interstitial water therefore builds up a phosphorus content far greater than that of the surrounding seawater.

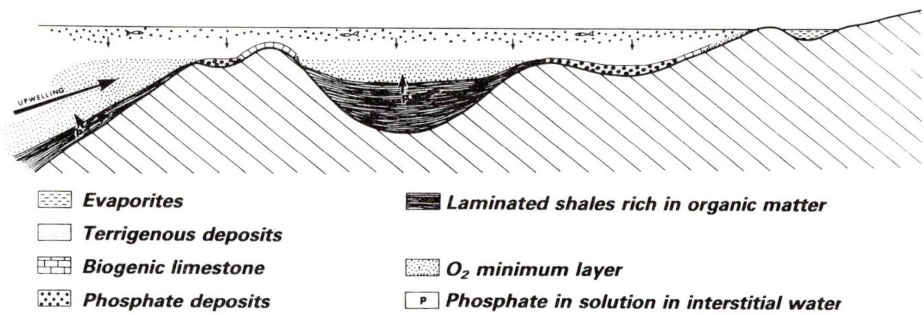

Fig. 39. *Summary of mechanisms of phosphate formation.*

3. Phosphate minerals precipitate from the interstitial water, near the water–sediment interface, and the margins, particularly the upper boundary, of the O_2 minimum layer. This environment, which is slightly oxidizing, encourages stability of the humic compounds formed from planktonic organic matter; the humic acids seem to play a direct role in phosphate precipitation and in the formation of phosphate particles.

4. Phosphate sedimentation as a whole is a slow phenomenon, for which periods or regions of rapid sedimentation are unsuitable. It requires a highly stable geological environment.

5. Phosphate deposits are very widely distributed along the stratigraphic column, from the Precambrian to the Recent but there are 'preferential' periods which appear to be associated with phases of increasing sea temperature, periods in which the seas rise and spread widely over well developed continental shelves, while plankton is highly diversified and abundant, and the O_2 minimum layer becomes thicker, favouring accumulations rich in organic matter.

The development of an anoxic environment favours the formation of 'laminated shales' or 'black shale' facies, mainly clayey–marly and cherty and eventually slightly evaporitic. In the most highly oxygenated portions of the basin, calcareous and siliceous benthic organogenic facies develop. Phosphate deposits occupy an intermediate position between these two facies, and are influenced by them to a greater or lesser extent.

6. Regions favourable to a large-scale phosphate deposition are those in which upwelling currents are the most efficient; an apparently important precondition is a latitude which our present knowledge suggests must be less than 40°. In such regions, sedimentation is also greatly condensed, terrigenous supply is slow and phosphate supply is higher because of the proximity of thick organic deposits in a reducing environment within which the phosphorus remains mobile.

7. The formation of a sufficiently high phosphate concentration to produce an ore deposit requires a generally shallow deposition of sediment and the existence of a trap favouring the accumulation and evolution of organic matter; the size of the ore deposit formed depends on that of the trap.

8. The surface weathering of phosphate deposits often leads to their natural enrichment. Intensive weathering, however, may lead to the formation of aluminium phosphorites. The secondary evolution of these deposits may also lead to their enrichment in uranium.

Chapter 6

Prospecting and mining of deposits

1. PROSPECTING METHODS

Prospecting techniques are direct applications of our knowledge of the geology of phosphate deposits. Prospecting generally takes place in three successive phases:

1. Determining the favourable geological periods for phosphate sedimentation.

2. Identification of zones favourable to the accumulation of phosphate deposits.

3. Detailed study of these zones.

1.1. Determining the favourable geological periods

Favourable geological periods may be determined directly from the age of deposits or phosphatic sequences already known in the region being prospected, in regions contiguous to it now or that were contiguous during the period in question, allowing for the relative movement of continents during geological time. Information may also be indirectly deduced from the general palaeogeographical and sedimentological characteristics of the phosphate deposits, summarized at the end of Chapter 5, together with the search for phosphate occurrences in the field or among the sample collections. This kind of work naturally assumes that the presence of phosphate can be recognized.

At the macroscopic level, phosphate rocks are not always easy to identify, which is why, for example, the outcropping phosphate occurrences of the Precambrian in India, the Infracambrian in western Niger, the Cambrian of the

Montagne Noire and the Silurian of the Vendée in France, the Devonian in Iran, the Jurassic and Senonian in Greece, the Senonian in Gabon, etc. were for a long time undiscovered by geologists travelling in these regions. The use of a molybdenum-based reagent may make identification easier: dilute nitric acid is poured onto the rock to be identified, followed by ammonium molybdate; a yellow precipitate forms in the presence of calcium phosphate[(1)]. This qualitive test is somewhat over-sensitive but has the advantage of being very quick. A 30 – 50% concentration of nitric acid is used, and a 10 – 15% concentration of the molybdate, to which 0.1 – 0.2% of ammonia is added, to make it easier to conserve.

With practice, it is possible to obtain an approximate quantitative estimate of the phosphorus content of a rock using this test; it is, however, wise to make a careful selection of samples to be sent to the laboratory, by using the semi-quantitative test developed by Shapiro (1952) at the U.S. Geological Survey. In this test, coloured reference tubes are compared with the colour of the phosphovanadomolybdate obtained after a given quantity of the sample to be tested has been treated with a solution of nitric and sulphuric acids and ammonium vanadomolybdate. Concentration is determined using about 3 mm^3 of powder, measured using a depression at the end of a wooden or metal stick; the technique requires great care by the operator if the test is to be representative and repeatable. The maximum error is about 5% and may be reduced to 2% if scales are used, but this makes the equipment less portable.

Phosphate rocks may often be identified also by their higher radioactivity compared with associated rocks. Modern portable scintillometers, which take up very little space, are being increasingly used in phosphate prospecting; the Middle Jurassic phosphatites of Epirus were discovered in this way, by teams from the Geological and Mining Institute of Greece. Note, however, that certain phosphate beds, even quite important ones, are no more radioactive than the surrounding rocks of the geological environment in which they occur (e.g. those of the Devonian system of Iran, and the Precambrian system of India).

A scintillometer can be used for phosphate prospecting in outcrops, wells and drillings. (Note that borehole casing does not mask radioactivity.) Gamma-ray logs are always very revealing.

Tests on airborne phosphate prospecting by scintillometers (using aeroplanes or helicopters) have also been carried out in various countries, e.g. in the northern part of the Paris Basin in 1961, in Algeria in 1961–1962 (Georgel *et*

---

[1] This does not apply to aluminium phosphates, which have to be treated with a hot solution of concentrated nitric and sulphuric acids.

al., 1962) and in Senegal in 1964. They have shown that only outcropping phosphate beds are likely to be revealed by this kind of test, as a barren cover of less than a metre thick is enough to cut off radioactive effects. Unusual radioactive behaviour can also be produced by nonphosphatic formations: e.g. marls rich in organic matter in Algeria, and laterites in Senegal. The necessity of flying at a low altitude (not above 100 m) parallel to the surface of the ground, also makes this kind of prospecting difficult over irregular terrain.

Geochemical phosphate prospecting can be considered for regions poor in outcrops. This method has been used in Benin by a U.N. mission, with encouraging results, as geochemical anomalies make it possible to trace phosphate beds accurately underneath the soils which cover them; however, such anomalies do not generally give a precise idea of the richness of the beds beneath. The method is difficult to apply in agricultural regions where phosphate fertilizers are used.

1.2. Identification of favourable zones

Deposits may be discovered directly by searching for exposed phosphate occurrences, as discussed above. However, this does depend on a combination of favourable circumstances, as in phosphate sedimentary basins of the type found in Senegal, for example, the area in which phosphate occurrences may be found may extend far beyond that covered by any deposit of economic value – the deposit, if there is one, may be at a considerable distance from the first occurrences discovered. In most cases, therefore, such traces only provide support for sedimentological and palaeogeographical reasoning concerning the identification of promising new rock stratigraphic units.

Once rock units have been selected, the choice of favourable zones is essentially a palaeogeographical problem consisting primarily of identifying regions devoid of any significant terrigenous supply, showing shallow condensed marine sedimentation, and which may contain traps. Palaeogeographical maps showing variations in thicknesses and facies of the terrain investigated are the usual instruments associated with this stage of prospecting. The classification of zones selected in this way, however, must also take economic criteria into account.

1.3. Detailed study of favourable zones

The third stage is systematic mining research, which itself consists of a number of phases, the first being the definition of economic criteria, in particular the minimum level of reserves and the maximum depth of the deposit for ore extraction to be viable. This makes it possible to decide on the spacing of exploratory works and on the type of drilling equipment required for the second phase, i.e. delimitation of the zone or zones likely to contain a deposit.

When the phosphate bed is radioactive and less than 50 – 70 m deep the use of compressed-air drills of the 'wagon drill' type, in conjunction with systematic radioactivity measurements to locate phosphate beds, allows rapid exploration at low cost. In recent prospecting work in the Senonian chalk of the northern part of the Paris Basin, these techniques allowed 3 – 4 pits between 30 and 40 m deep to be drilled daily with a single drill (Coulombeau, 1979).

Later phases will be to confirm the presence of ore deposits and define their characteristics systematically. The procedure adopted is standard, and there is no need to describe it in detail here. However, it is worth pointing out that there is a particular problem in taking core samples in phosphate ore, which is a soft rock liable to be encountered very suddenly, during drilling, after a very hard calcareous or cherty bed; generally, however, the use of gamma-ray logging is valuable in drilling for core-samples, and, where depth permits, pits are preferable to borings.

The final phase of this stage is an economic study that will primarily consider the possibility of producing a commercial concentrate transportable over long distances, i.e. with a high phosphate content. With the development of the phosphate industry during the last few decades leading to the increasingly widespread manufacture of finished products, particularly phosphoric acid and fertilizers, near the deposits themselves, it may be better to produce a less rich concentrate, which reduces operating costs and increases the mine's exploitable reserves. It is becoming increasingly difficult to separate the study of a particular ore deposit from overall considerations of all the activities that will develop downstream of it.

2. METHODS OF EXTRACTION AND PROCESSING

Between several hundred thousand and several million tonnes of crude ore are extracted annually from individual phosphate mines. There are underground and also open pit workings, the choice between these two methods depending on the regularity of the ore-bearing bed, the thickness of the overburden and its hardness, tectonic phenomena, etc. Deposits in Morocco, Tunisia, Egypt and the Rocky Mountains are worked partly underground, whereas in Florida, Tennessee, Senegal, Togo and Algeria, etc. only open cast mines are used.

A major advantage of open pit workings is that a number of phosphate beds can be mined simultaneously, leaving aside intermediate barren rock; such operations can, moreover, be highly mechanized. Very high capacity equipment is used for extraction: draglines with buckets whose capacity may exceed

50 m^3, with booms of up to 100 m long, bucket-wheel excavators capable of extracting up to 800 m^3/h, mechanical shovels with buckets of about 7 m^3 capable of extracting 1000 t/h. Transport inside the mine is provided by trucks with payloads of up to 100 t or even 225 t. The capacity of this equipment means that it is possible to work with up to 40 m of overburden and a stripping ratio (ratio of thickness of overburden to thickness of ore) greater than 5 and as high as 10 in favourable circumstances.

Underground workings have also spurred efforts towards mechanization in Morocco, Tunisia, the U.S.A. and the U.S.S.R., and great progress has been made in mechanical cutting equipment with movable hydraulic supports. Very large diameter drills (1.30 – 3.50 m) make it possible to sink wells from the surface or from a tunnel, or to work in circular tunnels. All these improvements have led to a noticeable reduction in underground costs.

The lowest acceptable phosphate content of an ore depends on a combination of factors, particularly on the difficulties of extracting and dressing it, the purpose for which the upgraded rock or concentrate is to be used, and on transport costs. Thanks to the progress that has taken place in extraction and beneficiation techniques, this minimum could be as low as about 10% P_2O_5 or even less (e.g. Florida); it is nevertheless generally higher, and the exporting countries in Africa and the Middle East exploit crude ores with a P_2O_5 content rarely below 25%. For local use the minimum grade of ore can easily be lower: about 15% for Tennessee ore converted on site and 10% for the phosphatic chalk in the Paris Basin.

The main purpose of ore dressing is to raise its P_2O_5 content to commercial standards, generally between 30 and 37% P_2O_5. Where necessary, ore may also be dressed to reduce the proportion of carbonates, 'feral' (Fe_2O_3 + Al_2O_3), organic matter, magnesium, chlorine and certain trace elements that may be harmful, such as cadmium, lead, chromium, arsenic, mercury and vanadium.

The most commonly adopted methods of dressing are drying followed by dust-removal (Djebel Onk, Christmas Island), sorting by grain size, either dry (Israel) or wet (Togo), calcination at 700 – 900°C (Rocky Mountains) to eliminate organic matter, or 850 – 900°C followed by washing to reduce the proportion of $CaCO_3$ (Djebel Onk, Morocco), flotation to eliminate silica (Senegal, Florida) and magnetic elimination of ferruginous particles (Togo).

The chlorine content can generally be reduced by washing in fresh water (Togo), and the cadmium content by calcination (Nauru). Magnesium in carbonates or silicates is generally not so difficult to eliminate, but this is not the case where magnesium occurs within the apatite lattice.

The processing of phosphate ores is still the subject of considerable research, particularly on the treatment of fine-grained particles (slimes) or ores with a carbonate gangue, which form the bulk of the reserves to be mined in the future. In the latter case, research is concentrating on improving calcination in order to maintain the ore's reactivity and reduce the energy consumption involved, and, above all, on flotation, which is less expensive.

Chapter 7

The phosphate industry

1. ECONOMIC DATA

The commercial concentrates produced in the world are essentially calcium phosphates (i.e. phosphatites), while aluminophosphatites, mined only in Senegal, represent not more than about 0.2% of total world output. The grade of a concentrate is evaluated in terms of percentage P_2O_5, or as the equivalent in tricalcium phosphate $Ca_3(PO_4)_2$, known as BPL (bone phosphate of lime) or TPL (triphosphate of lime) or sometimes TCP. Multiplication by 2.185 allows conversion from the P_2O_5 to the BPL figure, and 0.457 for the inverse conversion.

World output, estimated in millions of tonnes of commercial concentrate, increased from 3.5 in 1900 to 50 in 1963, 60 in 1965, nearly 124 in 1978, and over 139 in 1981 (Fig. 40). A reduction in consumption in numerous countries, however, left large stocks of unsold rock near the deposits in 1981, and in 1982 production was reduced to about the 1978 level. Table 16, based on statistics supplied by the IFA (International Fertilizer Industry Association), gives details of the quantities delivered by various producing countries, i.e. the quantities of commercial ore actually sold from 1980 to 1982, and also the average percentage P_2O_5 contents. This table shows that the U.S.A. is by far the biggest producing country, followed by the U.S.S.R. and Morocco; it also shows that 79% of deliveries (over the whole three-year period) are of sedimentary origin, 2.4% being guano derivatives and 18.6% of magmatic origin.

Morocco is the biggest exporter of ore, followed by the U.S.A. If, however, we take all forms of phosphorus into account, the U.S.A. is once more the largest, as it exports a far greater quantity of converted products: phosphoric and superphosphoric acid, triple superphosphate and, above all, ammonium phosphate (Table 17). The world market as a whole is, moreover, gradually moving towards an increase in exports of phosphoric acids and fertilizers at the expense of rock phosphate. In Western Europe, for example, the share of phosphate products represented by crude phosphate fell from 88% in 1977 to 83% in 1982, and 78% in 1983 according to preliminary estimates. Florida and North Carolina exported not more than 47% of the phosphate rock produced.

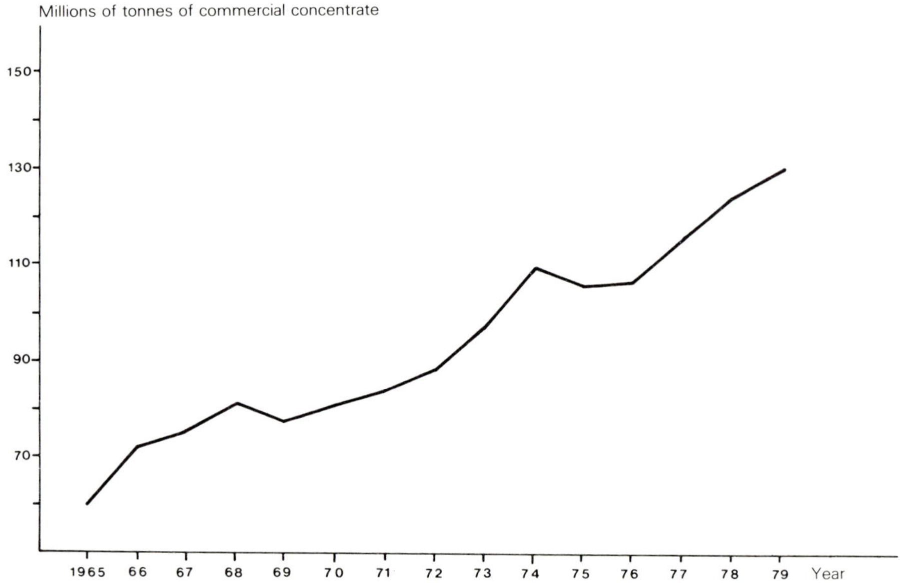

Fig. 40. *Changes in world production of commercial concentrates from 1965 to 1982.*

Table 16. *Deliveries and export of commercial concentrates in 1980 – 1982 (in millions of tonnes)*

•		1980 P$_2$O$_5$(%)	1980 Deliveries	1980 Exports	1981 P$_2$O$_5$(%)	1981 Deliveries	1981 Exports	1982 P$_2$O$_5$(%)	1982 Deliveries	1982 Exports
S	ALGERIA	30.99	994.9	768.4	30.58	935.3	791.1	31.05	948.4	703.7
S	AUSTRALIA				32.67	5		30.96	181.3	35.1
I	BRAZIL	35.56	2 921.3		34.65	2 471.5		34.63	2 769	
S	CHINA	32	6 700		22	11 500		25	12 500	
G	CHRISTMAS IS.	35.11	1 638.3	1 638.3	35.10	1 359.9	1 359.9	35.16	1 367	1 367
S	COLOMBIA	28.37	2.3		24.59	14.7		23.46	18.2	
S	EGYPT	27.89	574.4	257.8	28.15	577.9	229	27.81	684.1	290.6
I	FINLAND	36.37	124.8		36.37	200.9		37	231.5	
S	INDIA	33.23	461.7		33.22	470.2		31.72	552.4	
S	IRAQ							22		
S	ISRAEL	31.89	2 398	1 905.4	31.54	2 289.4	1 744	30.22	2 820.8	1 847.7
S	JORDAN	32.6	3 612.6	3 612.6	32.76	3 515.8	3 515.8	32.80	3 801.3	3 562.3
S	MEXICO	24.74	330		29.99	233.4		30.20	177.6	33.7
S	MOROCCO	32.23	18 865.5	16 457.1	32.10	19 139.9	15 635.5	31.99	17 703.7	13 976
G	NAURU	38.5	2 086.7	2 086.7	38.5	1 480	1 480	38.5	1 358.8	1 358.8
I	NORTH KOREA	30	500		30	500		32	500	
S	PERU	30.75	13.9		30.5	19.5	10.5	30.5	10	
I	SOUTH AFRICA	36.35	3 139.6	3.6	36.21	2 731.2	104.6	36.3	2 855.5	199.5
S	SENEGAL	35.91	1 434.5	1 377.8	36.14	1 208.1	1 155.1	35.97	1 259.4	1 202.9
I	SRI LANKA							34	15.1	
I	SWEDEN	38.9	77.6	60.8	38.57	112	94	38.09	145	130
S	SYRIA	30.46	861.7	823	30.41	1 069.8	936.5	30.66	928	722
S	TOGO	36.17	2 898.2	2 896	36.41	2 219.7	2 215.4	36.30	2 056.6	2 055.1
S	TUNISIA	29.26	4 527	1 449.1	28.91	4 204.5	1 046	28.92	4 451.7	1 139.4
S	TURKEY	29.5	21.3		29.5	43		29.5	26.3	
S	U.S.A.	31.21	55 539.4	14 381.9	30.96	47 755	10 499.9	31.07	39 511	9 933 4
I&S	U.S.S.R.	35.31	26 100	4 558	35.61	25 200	5 019.9	35.35	26 600	5 226.8
S	VIETNAM	35	400		35	500		35	500	
I	ZIMBABWE	35.5	130		35	125		34.7	120	
	Total		136 353.8	52 276.6		129 881.4	45 837		124 092.8	43 784
	Mean content	32.69			31.71			32.04		

Nature of the ore: S = sedimentary; G = guano; I = igneous.

Table 17. *Nature of exports from the main phosphate-producing countries.*

	Ore (10^3 tonnes)			Phosphoric acid (tonnes P_2O_5)			Triple superphosphate			Ammonium phosphates (tonnes P_2O_5)		
	1977	1978	1979	1977	1978	1979	1977	1978	1979	1977	1978	1979
Algeria	698	619	859									
Egypt	220	231	215									
Israel	968	1 425	1 720									
Jordan	1 787	2 159	2 727									
Morocco	15 816	17 746	18 006	220 585	243 915	359 394	81 701	84 557	77 799	13 844	25 751	6 868
Senegal	1 832	1 758	1 601									
Togo	2 885	2 849	2 990									
Tunisia	1 897	1 667	1 516	174 916	175 440	196 505	205 277	208 988	234 249			
U.S.A.	13 230	13 414	14 673	424 119	443 000	762 469	504 638	677 393	663 824	1 365 337	1 961 545	2 073 955
U.S.S.R.	4 242	3 992	4 130									
Others	4 887	5 657	5 006									

The price of phosphate concentrates has varied considerably since the beginning of the 1970s. For example, the price per tonne of Moroccan concentrate, 75 BPL in Casablanca (f.o.b. price) increased from \$11.75 in 1972 to \$14.20 in 1973, \$42 in January 1974 and \$68 in January 1975. This price increase led to a considerable fall in demand, which in turn triggered a fall in prices. The recovery that followed led to an official (f.o.b.) price for Moroccan 75 BPL of \$42 in 1979, but the depressed state of the market between 1981 and 1983 led to further falls and to unstable prices.

2. WORLD RESOURCES AND RESERVES

The varied nature of the data available makes it very difficult to evaluate the world's phosphate resources. Not all deposits are equally well known, and the P_2O_5 content of different ores can vary widely (from 4.5% for the Mexican deposit at Santa Domingo, Baja California, to 35% in certain parts of the Moroccan deposits) and the concept of 'reserves' is not the same in all of them.

An attempt was made in 1981, by the U.S. Geological Survey to standardize definitions of reserves and resources by proposing an interesting general classification of ore resources and reserves (U.S.G.S. Circular 831, 1981). Its application to phosphate ores is recommended. The resource/reserve categories are:

1. Phosphate *resources* are all rocks from which profitable industrial phosphate production can be envisaged now or in the foreseeable future, taking into account relevant economic and technological factors.

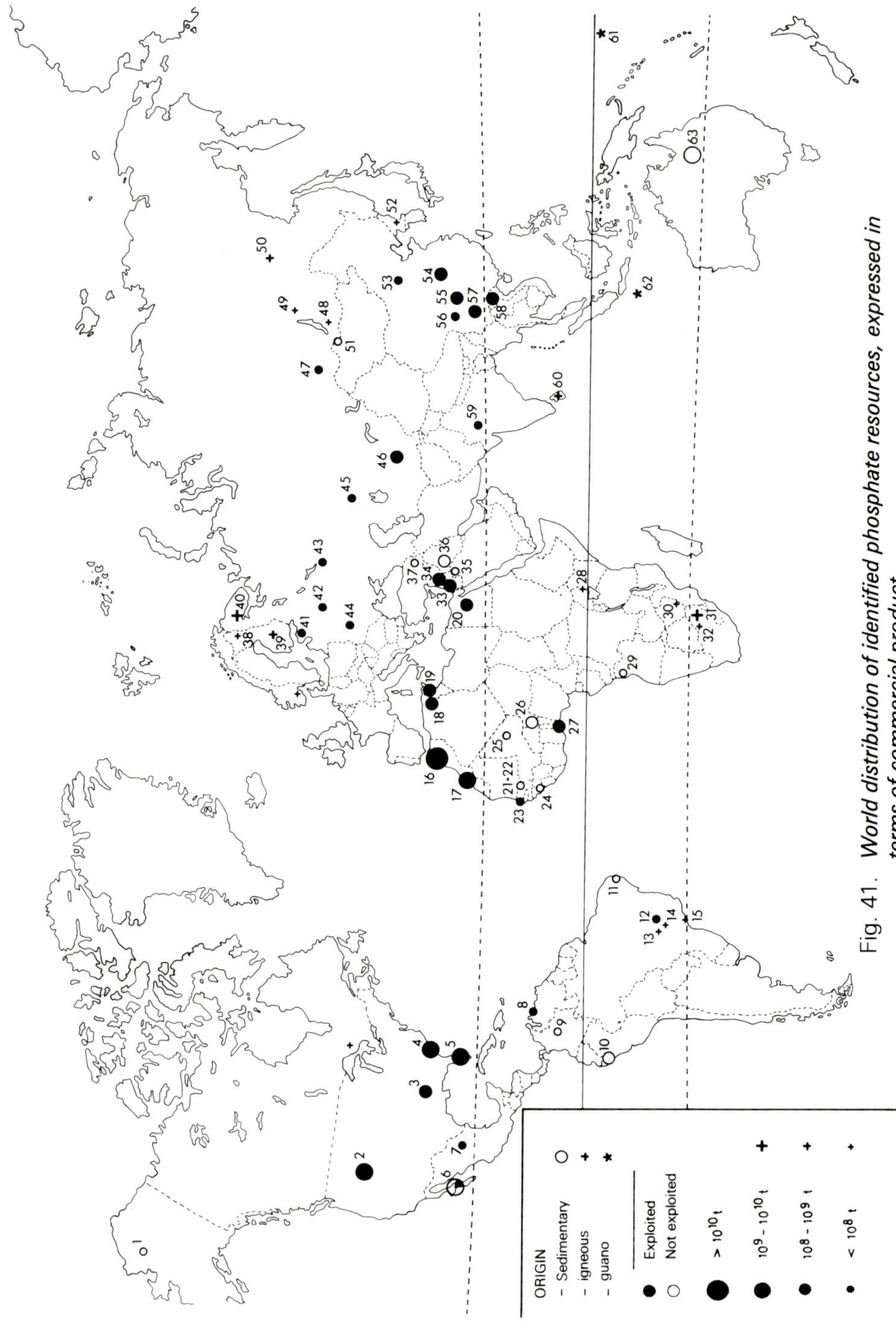

Fig. 41. *World distribution of identified phosphate resources, expressed in terms of commercial product.*

Resources can be evaluated in tonnes of ore *in situ*, tonnes of commerial product derivable from it, or in tonnes of P_2O_5 *in situ* or recoverable.

2. *Identified resources* are those that have been the subject of specific prospecting operations or that are an obvious extension of already prospected resources. Resources, the presence of which has been deduced from geological observation and reasoning, but which have not yet been identified are termed 'undiscovered resources' by the U.S.G.S.; the author would term these *'possible resources'*.

3. Phosphate *reserves* are identified resources, the mining of which seems profitable or, in some cases, almost profitable ('marginal reserves') under present economic conditions. Identified resources that cannot be classified as 'reserves' are *'sub-economic resources'*.

The system proposed by the U.S.G.S. distinguishes between three categories of reserves ('measured', 'indicated' and 'inferred'), depending on how much is known about them. Advances in the use of geostatistical methods in the evaluation of the deposits, however, make it increasingly easy to characterize reserves by the degree of certainty of what is known about them.

1–5 U.S.A.: 1 Alaska; 2 Idaho, Wyoming, Montana; 3 Tennessee; 4 North Carolina; 5 Florida. *6–7 Mexico*: 6 Baja California; 7 Nuevo Leon, Queretaro, etc. *8 Venezuela*: Falcon (Riecito). *9 Colombia*: Norte de Santander, Santander, Boyaca, Huila. *10 Peru*: Sechura (Boyovar). *11–15 Brazil:* 11 Pernambuco (Olinda); 12 Minas Gerais (Patos de Minas); 13 Goias (Catalao); 14 Minas Gerais (Tapira, Araxa, etc.); 15 Sao Paulo (Jacupiranga). *16–17 Morocco*; 16 Oulad Abdoun, Ganntour; 17 Bou Craa. *18 Algeria*: Djebel Onk. *19 Tunisia*: Gafsa Basin, Sra Ouartane. *20 Egypt*: Red Sea, Nile Valley, Abu Tartur. *21 Mauritania*: Bofal. *22–23 Senegal*: 22 Matam; 23 Taïba. *24 Guinea Bissau*: Farim. *25 Mali*: Tilemsi. *26 Niger, Bukina Faso (Upper Volta), Benin*: Tapoa, Kodjari, Mekrou. *27 Togo*: Hahotoe. *28 Uganda*: Sukulu. *29 Angola*: Cabinda, Zaire. *30 Zimbabwe. 31–32 South Africa*: 31 Palabora; 32 Glenover. *33 Israel – Jordan*: Negev Desert — Ruseifa, Al Hasa. *34 Syria*: Khneifiss, Eastern. *35 Saudi Arabia*: Thaniyat, Turayf. *36 Iraq*: Akashat. *37 Turkey*: Mazidagi. *38–39 Finland*: 38 Sokli; 39 Siilinjärvi. *40–50 U.S.S.R.*: 40 Khibiny, Kovdor; 41 Maardu; 42 Moscow Basin; 43 Rudnichny; 44 Byelorussia, Ukraine; 45 Aktyubinsk; 46 Karatau; 47 Abakan; 48 Oshurkov; 49 Synnyr; 50 Seligdar. *51 Mongolia*: Hobsogol. *52 North Korea. 53–57 China*: Fanskan, Chingshan, Kaiyang, Omei, Kunming–Kunyang; *58 Vietnam*: Lao Kai. *59 India*: Udaipur. *60 Sri Lanka*: Eppawella. *61 Nauru. 62–63 Australia*: 62 Christmas Island; 63 Queensland (Duchess).

According to the U.S. Bureau of Mines (1983), the total resources identified in the world, evaluated in terms of commercial product (i.e. in terms of apatite concentrates resulting from mining and benefication) amount to 105×10^9 t, of which 55×10^9 t are classed as reserves. These quantities respectively represent 500 and over 200 times the present annual consumption, so that the known deposits seem capable of covering world demand for some considerable time to come. These resources do not include the phosphate deposits of the present ocean floors, whose economic potential is still not certain.

Total identified resources, however, are unevenly distributed. About 76% of them are in Africa, particularly in Morocco (nearly 60% of the world's resources), and 12% in the U.S.A. Resources in Western Europe, a big phosphate consumer, on the other hand, are very limited, and those of South America and South-east Asia cannot meet the regions' own needs. A serious problem in many countries is the lack of foreign currency with which to import the additional phosphate needed. The search for new deposits should therefore continue, despite the abundance of known resources.

Figure 41 uses various sources in information to give the order of magnitude of the identified resources in terms of commerical productes in the main deposits or groups of deposits around the world.

3. USES OF PHOSPHATES

In 1971, Dumon gave the following figures for the various proportions of phosphate used for different purposes:

- 77% for fertilizers obtained by reacting phosphate rock with acid;
- 4% for fertilizers applied directly (ground, natural phosphates);
- 8% for thermal (elemental) phosphorus and phosphoric acid;
- 6% for cattle feed;
- 5% for polyphosphates.

3.1. Manufacture of fertilizers by acid treatment

Sulphuric acid is generally used, producing:

- superphosphate $Ca(H_2PO_4)_2H_2O + CaSO_4$, containing 15 – 20% P_2O_5;
- 'wet process' phosphoric acid, which is then used to produce:
  - triple superphosphate $Ca(H_2PO_4)_2$, containing 40 – 45% P_2O_5;

- monoammonium phosphate $NH_4H_2PO_4$, containing 60% P_2O_5 + 12% N;
- diammonium phosphate $(NH_4)_2HPO_4$, containing 53% P_2O_5 + 21% N.

Superphosphate, usually called 'single superphosphate' (SSP), is the oldest of these products. Its use has become far less common, but nevertheless continues. As its phosphorus content is less than that of phosphate rock concentrates, it is mainly produced close to the areas where it is actually used. In the manufacture of superphosphate, higher proportions of 'feral' and magnesium can be tolerated than in the case of phosphoric acid.

'Wet process' phosphoric acid is the most important intermediate product in the phosphate industry. Originally confined to the manufacture of fertilizers, it is being increasingly used for other purposes for which phosphate products are required. The acid produced by treatment with sulphuric acid initially contains 30 – 32% P_2O_5; after concentration and purification, the proportion may reach 55 – 60%. Further concentration may lead to 'superphosphoric' acid, containing up to 75% P_2O_5.

This process requires very high quality phosphate rock concentrates, which must be reasonably reactive, have a CaO/P_2O_5 ratio less than 1.6, and contain less than 0.5% MgO, 3 – 4% Fe_2O_3 + Al_2O_3, 0.03% Cl, and very little organic matter. Slightly higher proportions of CaO, MgO or Cl, while tolerable, increase production costs and the cost of plant. A silica content of over 3–5% leads to a number of disadvantages; some silica must, however, be present to fix fluorine and extract it in the form of fluorosilicic acid, H_2SiF_6, which can then be used for the production of the fluorosilicates and fluorides of potassium, sodium, aluminium, or even cryolite.

A further by-product of the manufacture of phosphoric acid by the wet process is phosphogypsum, an impure gypsum of which 4.5 t is obtained per tonne of P_2O_5. No satisfactory way of using all of it has been found yet, and for the moment it is basically a material that is hard to get rid of and occurs as waste in very large quantities.

The manufacture of phosphoric acid by the wet process also makes it possible to extract most of the uranium contained in the phosphate rock used; most processes use organic solvents in dilute phosphoric acid. Falling uranium prices over the past few years have, nevertheless, put a brake on work in this sphere of production.

Triple superphosphate (TSP) and, above all, monoammonium and diammonium phosphates (MAP and DAP) are the most widely manufactered fertilizers today. As the P_2O_5 content is greater than that found in concentrates, their manufacture is being gradually undertaken in those countries that mine the deposits.

3.2. Fertilizers for direct application (ground phosphates)

These are phosphate ores ground to about 50 μm and applied directly to the soil, either alone or mixed with other fertilizers. Unlike those discussed above, they are not water soluble, and their quality is defined by their solubility in a 2% formic acid solution. European standards stipulate a minimum solubility of 55%. As has been shown by Lehr *et al.* (1967), solubility increases with the degree of substitution of PO_4 by CO_3 in the apatite lattice. The ore at Gafsa (75% solubility) is, therefore, more suitable for the production of ground phosphates than those of Morocco (59%) or Senegal (16%). Apatites of magmatic (igneous) origin are the least soluble.

These fertilizers are very suitable for highly leached soils and for acid soils in general. They have the advantage in that they can be produced from mediocre ores, requiring only a small amount of energy; consumption of ground phosphate is therefore constantly growing. It is also a good subsitute product for those agricultural regions accustomed to using basic slag produced by dephosphorization of iron ore, whose production has fallen considerably with technological changes in the iron industry.

The solubility of natural phosphates may also be increased by partial acidulation with sulphuric acid or possibly phosphoric or nitric acid. In addition to formic acid solubility, there is a water solubility level which depends on the quantity of acid added. These partially acidulated fertilizers, which were common during the last World War, are tending to become more widely used.

3.3. Other phosphate fertilizers

Among the phosphate fertilizers with good citric acid solubility, sintered phosphates and fused phosphates deserve special mention, even though world output levels are low, as it may be possible to manufacture them from ores for which it is difficult to find any other use.

For sintered phosphates, the Rhenania process is most widely used. This involves calcination, at 1100 –1200°C, of a mixture of ore, sodium carbonate and silica, in the presence of steam, to produce a phosphate of calcium and sodium with good citric acid solubility. A product of this type was manufactured at the B.R.G.M. (Baudet and Morizot, 1978) from a calcitic glauconitic ore; the resulting solubility was good not only in terms of phosphorus, but also in respect of the potassium from glauconite.

Fused phosphates require higher temperatures (1700°C) but treatments may not last as long. The Thermophos process is based on a reaction between serpentine (a ferromagnesium silicate) and crude ore, and produces a phosphate fertilizer containing magnesium which appears to give good results with all soils. High magnesium content in the ore apparently presents no obstacle in this case.

Calcination at 600°C, followed by crushing, also considerably improves the solubility of aluminium phosphorites. Fertilizer obtained in this way from Senegal (Thiès) phosphorite is called phospal, which is suitable for various soils, especially calcareous soil.

3.4. Thermal phosphorus and phosphoric acid

Elemental phosphorus is manufactured by thermoreduction of apatite phosphate in an electric furnace, using coke in the presence of silica. The electrical energy required for the furnace is nearly 15 000 kW per tonne of phosphorus produced. The resulting reactions lead to a release of carbon monoxide and phosphorus vapour and to the production of calcium silicate slag and ferrophosphorus. Phosphoric acid produced by the combustion of phosphorus in the air, followed by hydration, is extremely pure, which is the main reason for using this process.

Thermal phosphoric acid is used in the detergent industry, particularly in regions where the cost of electrical energy make is feasible (U.S.A., U.S.S.R., Canada). The use of phosphorus and acid produced in this way is more common in metallurgy, for various surface treatments of aluminium, copper and their alloys, and the nickel-coating of iron and steel. The food, pharmaceutical and petroleum industries (catalysts, special lubricants, additives, refining) also use them, and they also appear in numerous other applications, such as the manufacture of insecticides and pesticides, the textile industry (bleaching agents), water treatment, plastics, photography, etc.

With a higer cost price than wet-process phosphoric acid (the price difference in France in 1974 was about 50%), thermal acid is not in general use in the fertilizer industry, and it does not seem possible to use it for the treatment of imported phosphate concentrate. Since, however, the thermal process is far less sensitive to phosphate concentration levels and to impurities in the ore, as seen from Table 18 (certain ores in an electric furnace in Idaho have a P_2O_5 content of as little as 20%), it may in certain cases be the best method if local resources are to be used.

3.5. Cattle feed

Phosphatites used for cattle feed are calcined and defluorinated. The low proportions of fluorine that can be accepted mean that ores naturally poor in

Table 18. *Composition of ores used in electric furnaces in Tennessee and Idaho* (from Hurst, 1958, p. 1176).

	Tennessee	Idaho	
	Nodules %	Conglomerate %	Nodules %
P_2O_5	25–28	25–27.5	23.5–27
CaO	34–39	35–38.5	37.5–39.5
SiO_2	24–27.5	20–26	17–26
Fe_2O_3	2.5–4.5	4.1–4.5	1.2–1.8
Al_2O_3	4–6.5	5–6.5	3.2–6.5
MgO	0.4	0.3–0.5	0.5–2
F	2.3–2.5	2–2.5	2–2.5
Alcali metals	0.8–1.3	0.9–1.5	1.8
MnO	0.3	0.3	0.12–0.15
TiO_2	0.1	0.1	

this element are at a premium in this industry; this has been the case, for example, with Curacao ore, in the Caribbean. That part of the alumimium phosphorite output in Senegal which is low in fluorine is aimed at this market, under the name of 'Polyfos'.

3.6. Polyphosphates

Polyphosphates are being manufactured to an ever-increasing extent from purified wet-process phosphoric acid. The type most widely used is sodium tripolyphosphate, which is an important component in domestic and industrial washing agents, because of its emulsifying capacity and ability to keep salts of the alkaline earths and heavy metals in solution. There is a fear that polyphosphates in detergents cause serious pollution, affecting the taste of beverages and, above all, contributing to the eutrophication of rivers, lakes and drainage basins. The trend today, therefore, is to limit their use. Various experts, however, argue that these substances are less polluting, with respect to many others, than has been suggested.

Chapter 8

Conclusion

Studies carried out over the last few decades, giving rise particulary to the multidisciplinary approach to the problem, have allowed us to form far clearer ideas concerning the mechanisms of phosphate genesis and of the formation of sedimentary phosphate deposits. Our knowledge is still not complete, however, and further research remains to be done, particularly to assist prospecting in regions that are currently deficient in resources of suitable size and quantity, and to provided new ways of making profitable use of the ores.

The large deposits discovered over the past twenty years in India (Rajasthan), Australia (Queensland), Brazil (Minas Gerais) and West Africa (Bou Craa, eastern Senegal, Burkina Faso, Niger) prove that it is still possible to discover phosphate accumulations of economic importance, but inevitably prospecting will become increasingly difficult. Prospecting equipment has to be refined further, therefore, and existing deposits, including those on the continental margins, have to be closely studied in order to obtain a clearer idea of the palaeogeographical and sedimentological factors favourable to phosphatite formation. In this connection, organic matter and organisms, including bacteria, need to be carefully studied.

The increase in world consumption has also made it necessary to mine lower-quality ores, as is made clear by the decrease in average P_2O_5 content from 33.30% (in 1971) to 31.90% (in 1982) of Moroccan commercial ore (IFA statistics). At the same time, the beneficiation of ores is becoming increasingly important, particularly those with a carbonate gangue, which form a large proportion of identified phosphate resources. There is, therefore, a tendency to try to improve the economic potential of ores by extracting by-products (fluorine, uranium, rare earths, vanadium, etc.) and by eliminating harmful elements such as cadmium, which has attracted considerable attention over the past few years. Cadmium, whose proportion varies from less than 10 ppm to 80 – 90 ppm in commercial phosphate concentrates of sedimentary origin, is liable to contribute to soil pollution by way of fertilizers, or to water pollution when phosphogypsum is discharged.

The beneficiation and use of mined phosphate rock require detailed petrographic, mineralogical and geochemical data in order to be as efficient as possible. Obtaining such data is doubtless made easier by the variety and accuracy of results obtained in modern laboratories. It does, however, raise the issue of whether the observations are representative, i.e. whether the results often obtained, for various practical reasons, from a limited number of small specimens can be applied generally; this is the case, for example, with studies conducted using optical or electronic microscopes, with the Castaing microprobe or by X-ray diffraction measurements.

Data obtained from such investigations are often difficult to interpret, owing to the insufficiency of our present knowledge of the laws governing the relations between the physical or chemical characteristics of an ore and the way in which the initial deposit forms, its later evolution, the different fractions making up the ore and phosphate particles, as well as the nature and origin of these particles, etc.

To cover the demand for phosphate rock and to make the best possible use of identified resources, considerable and varied research into methodology is therefore required. It is again necessary to stress that such research will be all the more effective, the more consistently it is approached on a multidisciplinary basis, and the more closely the various international experts concerned cooperate with one another.

Bibliography

AL MALEH A. (1974): Les phosphates sénoniens du Nord-Ouest syrien; *C.R. Acad. Sci., Ser. D,* **278,** 1003-1006.

ALTSCHULER Z.S. (1973): The weathering of phosphate deposits. Geochemical and environmental aspects; in: *Environmental Phosphorus Handbook,* 33-96, Wiley, New York.

ALTSCHULER Z.S. (1980): The geochemistry of trace elements in marine phosphorites. Part I. Characteristic abundances and enrichment; *SEPM Spec. Publ.,* no. 29, 19-30.

ALTSCHULER Z.S., CLARK R.S. and YOUNG E.J. (1958): Geochemistry of uranium in apatite and phosphorite; *U.S. Geol. Surv. Prof. Pap.,* **314-D,** 45-90.

ALTSCHULER Z.S., CATHCART J.B. and YOUNG E.J. (1964): The Geology and Geochemistry of the Bone Valley Formation and its Phosphate Deposits, West Central Florida. A Guidebook for Field Trip No. 6; *Geol. Soc. Am. Conv.,* Nov. 1964. 68 pp.

ALTSCHULER Z.S., CUTTITTA F. and BERMAN S. (1967): Rare earths in phosphorites, – geochemistry and economic potential; *U.S. Geol. Surv. Prof. Pap.,* **575-B,** B1-B9.

AMES L.L. (1959): The genesis of carbonate apatites; *Econ. Geol.,* **54,** 829-841.

AMIT O. and BEIN A. (1982): Organic matter in Senonian phosphorites from Israel. Origin and diagenesis; *Chem. Geol.,* **37,** 277-287.

ANDERSEN J.P. (1975): Influence of pH on release of phosphorus from lake sediments; *Arch. Hydrobiol.,* **76** (4), 411-419.

ARDA O., SALTOGLU T., ALPARSLAN E. and AKYUZ T. (1976): Geochemical and mineralogical analyses of the Mazidagi phosphates containing uranium, vanadium, fluorine and other trace elements; and views on the probability of uranium recovery; *Bull. Miner. Res. Explor. Inst. Turkey,* **87,** 33-48.

ARMSTRONG F.A.J. (1965): Phosphorus; in: *Chemical Oceanography,* **1,** 323-364, Academic Press, London and New York.

ARTHUR M.A. and JENKYNS H.C. (1981): Phosphorites and palaeoceanography; *Oceanologica Acta,* Special Issue, Proceedings of 26th International Geological Congress, Paris, France. Oceanic Geology Colloquium, 1980, 83-96.

ARVIEU J.C. and BOUVIER O. (1974): Les processus chimiques de l'évolution des phosphates en sols calcaires; *Bull. Assoc. Fr. Etude Sol,* no. 4, 207-224.

ATLAS E. and PYTKOWICZ R.M. (1977): Solubility behaviour of apatites in seawater; *Limnol. Oceanogr.,* **22** (2), 290-300.

AUBRY M.P. (1972): Recherches pétrographiques, stratigraphiques et paléosédimentologiques sur les craies de Haute-Normandie; *Thèse de 3ème cycle,* Université de Paris VI.

BANERJEE D.M. (1971): Precambrian stromatolitic phosphorite of Udaipur, Rajasthan, India; *Geol. Soc. Am. Bull.,* **82,** 2319-2330.

BANERJEE D.M. (1978): Chemical rhythmicity in the Precambrian laminated phosphatic stromatolites and its bearing on the origin of algal phosphorite; *Indian J. Earth Sci.,* **5** (1), 102-110.

BARRIE J. (1967): The Geology of Christmas Island; *Rec. No.* 1967/37, *Bur. Miner. Resour. Geol. Geophys. Aust.,* 46 pp.

BASHIR SALEH (1977): Radioactivité et état d'équilibre radioactif de quelques phosphates jordaniens. Leurs caractères pétrologiques, minéralogiques, cristallochimiques et géochimiques; *Thèse de docteur ingénieur,* Institut National Polytechnique de Lorraine, Ecole Nationale Supérieure de Géol. Appliquée et de Prospection Minière, Nancy.

BATHURST R.G.C. (1971): Carbonate sediments and their diagenesis, in: *Developments in Sedimentology,* **12,** 620 pp., Elsevier, Amsterdam, Oxford and New York.

BATURIN G.N. (1969): Authigenic phosphate concretions in recent sediments of the South-West African shelf; *Acad. Sci. USSR, Earth Sci. Sect.,* **189,** 227-230.

BATURIN G.N. (1970): Recent authigenic phosphorite formation on the South-West African shelf; in: *The Geology of the East Atlantic Continental Margin,* ICSU/SCOR Working Party 31 Symposium, Cambridge, *Inst. Geol. Sci.* Report 70/13, 88-97.

BATURIN G.N. (1971): Formation of phosphate sediments and water dynamics; *Oceanology,* **11** (3), 372-376.

BATURIN G.N. (1972): Phosphorus in interstitial waters of sediments of the southeastern Atlantic; *Oceanology,* **12** (6), 849-855.

BATURIN G.N. (1974a): On the geological consequences of mass mortality of fish in the ocean;*Oceanology,* **14** (1), 101-105.

BATURIN G.N. (1974b): New data on the Upper Quaternary phosphorite concretions on the continental shelf of Southwest Africa;*Oceanology,* **14** (6), 1040-1044.

BATURIN G.N. (1982): Phosphorites on the sea floor. Origin, composition and distribution; in: *Developments in Sedimentology,* **33,** 343 pp., Elsevier, Amsterdam, Oxford and New York.

BATURIN G.N. and DUBINCHUK U.T. (1974): Microstructures of Agulhas Bank phosporites; *Mar. Geol.,* **16** (1), 1163-1170.

BATURIN G.N. and KOCHENOV A.K. (1974): Uranium content of oceanic phosphorites; *Lithol. Miner. Resour.,* **9** (1), 99-103.

BATURIN G.N and SHISHKINA O.V. (1973): Behavior of fluorine during phosphorite formation in the ocean; *Oceanology,* **13** (4), 523-527.

BATURIN G.N., BLISKOVSKIY V.Z. and MINEYEV D.A. (1972): Rare earths in phosphorite samples from the ocean floor; *Dokl. Akad. Nauk. SSSR,* **207** (4), 954-957.

BATURIN G.N., BLISKOVSKIY V.Z. and LISITSYN A.P. (1976): Upper Quaternary Phosphorite Nodules off the coast of Peru;*Oceanology,* **15** (13), 331-333.

BATURIN G.N., SHUMENKO S.I. and DUBINCHUK V.T. (1977): Coccolithophorids in phosphorites from seamounts in the North-West Pacific; *Oceanology,* **17** (1), 55-57.

BAUDET G. and MORIZOT G. (1978): Essais de grillages alcalins de phosphates français; *Comm. Congr. Soc. Ind. Minér., Le Touquet,* 1–2 June 1978.

BEERS J.R. (1964): Ammonia and inorganic phosphorus excretion by the planktonic Chaetognath, Sagitta hispida; *Conn. J. Cons. Int.* Explor. Mer, **29** (2), 123-129.

BELAYOUNI H. (1978): Etude des composés aminés contenus dans quelques séries phosphatées; *Thèse de 3ème cycle,* Université d'Orléans.

BELAYOUNI H. (1984): Etude de la matière organique dans la série phosphatée du bassin de Gafsa. Metlaoui (Tunisie). Application à la compréhension des mécanismes de la phosphatogenèse; *Doc. B.R.G.M.,* no. 77, 240 pp.

BELAYOUNI H. and TRICHET J. (1980): Contribution à la connaissance de la matière organique du bassin phosphaté de Gafsa. Informations fournies par l'analyse du potentiel pétrolier. In: *Géologie comparée des gisements de phosphates et de pétrole; Doc. B.R.G.M.,* no. 24, 37-59.

BELAYOUNI H., FAUCONNIER D., SLANSKY M. and TRICHET J. (1982): Etude du contenu organique des dépôts phosphatés du bassin de Gafsa (Tunisie); *Doc. B.R.G.M.,* no. 35, 70 pp.

BELINKO (de) G. (1977): Réflexions sur la genèse des gisements sédimentaires marins des phosphates; *Ind. Minér. (St-Etienne, Fr.),* **59** (6), 290-310.

BENTOR Y.K. (1980): Phosphorites. The unsolved problems; *SEPM Spec. Publ.,* no. 29, 3-18.

BERGE J.W. (1972): Physical and chemical factors in the formation of marine apatite; *Econ. Geol.*, **67**, 824-827.

BERNARD F. (1953): Rôle des flagellés calcaires dans la fertilité et la sédimentation en mer profonde; *Deep-Sea Res.*, no. 1, 34-46.

BERZINA I.G., KARPOVA M.I. and KHARITONOVA R.SH. (1976): Uranium distribution in nodular platform phosphorites; *Lithol. Miner. Resour.*, **11** (2), 215-222.

BIDAUT H. (1953): Notes préliminaires sur un mode de formation possible des phosphates dinantiens des Pyrénées; *C.R. 19th Congr. Geol. Int., Algiers,* **XI,** 185-190.

BIGOTTE G. and BONIPAS G. (1968): Faits nouveaux sur la géologie de la région de Bakouma; *Chron. Mines. Rech. Minière.*, **370**, 43-46.

BIZARD CH. (1953): Précision sur les caractères de sédimentation des niveaux phosphatés et leur relation avec la radioactivité dans le gisement des Ouled Abdoun (Maroc central); *C.R. Acad. Sci.,* **236,** 1587-1589.

BONEL G. (1972): Contribution à l'étude de la carbonation des apatites; *Ann. Chim. (Paris),* **7,** 65-88, 127-144.

BONEL G., LABARTHE J.C. and VIGNOLES C. (1975): Contribution à l'étude structurale des apatites carbonatées de type B; in: *Physico-chimie et cristallographie des apatites d'intérêt biologique,* 117-125, C.N.R.S., Paris.

BOUJO A. (1976): Contribution à l'étude géologique du gisement de phosphate crétacé ocène des Ganntour (Maroc occidental); *Sci. Géol. Bull.,* **43,** 227 pp.; *Thèse d'Etat,* Université de Strasbourg, 1972.

BOUJO A. (1980): Un modèle de sédimentation phosphatée contrôlée par le fond: le golfe septentrional marocain (Oulad Abdoun-Ganntour); in: *Géologie comparée des gisements de phosphates et de pétrole*; *Doc. B.R.G.M.,* no. 24, 129-140.

BOUJO A., and OULD JIDDOU EL H. (1983): Une découverte récente de phosphate: les gisements de Bofal et de Loubboïra (Mauritanie méridionale, région du fleuve Sénégal); *Chron. Mines Rech. Minière,* no. 472, 37-52.

BOUJO A., FAYE B., GIOT D., LUCAS J., MANIVIT H., MONCIARDINI C. and PREVOT L. (1980): The early Eocene of the Lake of Guiers (Western Senegal). Reflections on some characteristics of phosphate sedimentation in Senegal; *SEPM Spec. Publ.,* no. 29, 207-213.

BOUYSSE PH. (1969): Les phosphorites sous-marines; *B.R.G.M. Rep.* 69 SGL 204 GEO.

BOWEN H.J.M. (1966): *Trace Elements in Biochemistry*; Academic Press, London.

BRAMLETTE M.N. (1958): Significance of coccolithophorids in calcium-carbonate deposition; *Geol. Soc. Am. Bull.,* **69** (1), 121-126.

BRASSENS A. (1975): Mise en évidence de deux types d'eau et de deux localisations des ions $CO^{2-}/_3$ dans la matière phosphatée d'un phosphate sédimentaire marocain; *C.R. Acad. Sci.*, **280,** 879-881.

BRIDEN J.C., DREWRY G.E. and SMITH A.G. (1974): Phanerozoic equal-area world maps; *J. Geol.*, **82,** 555-574.

BRIDGE P.J. (1973): Guano minerals from Murra-el-elevyn Cave – Western Australia; *Mineral. Mag.*, **39,** 467-469.

BRITISH SULPHUR CORPORATION LIMITED: *Phosphorus & Potassium;* (Bimonthly plus 2 Annual Statistical Supplements), London.

BRITISH SULPHUR CORPORATION LIMITED (1980): *World Survey of Phosphate Deposits,* 4th edn, M.C. Mew (Ed.), London, 238 pp.

BRONGERSMA-SANDERS M. (1957): Mass mortality in the sea; *Geol. Soc. Am., Mem.*, **67** (1), 941-1010.

BRONGERSMA-SANDERS M. (1966a): The fertility of the sea and its bearing on the origin of oil; *Adv. Sci.*, **23,** 41-46.

BRONGERSMA-SANDERS M. (1966b): Barium in pelagic sediments and in diatoms; *Koninkl. Ned. Akad. Wetensch., Proc. B,* 93-99.

BROPHY G.P. and NASH J.T. (1968): Compositional infrared X-ray analysis of fossil bone; *Am. Mineral.*, **53,** 445-454.

BRUNEAU L., JERLOV N.G. and KOCZY F.F. (1953): "Physical and Chemical Methods, Appendix, Table 1, Physical and Chemical Data"; in: *Swedish Deep-Sea Expendition Reports* 1947-48. no. 4, 1953, 99–112.

BURNETT W.C. (1974): Phosphorite deposits from the sea floor off Peru and Chile; *Radiochemical and Geochemical Investigations Concerning their Origin,* HIG 74-3, Hawaii Institute of Geophysics, University of Hawaii.

BURNETT W.C. (1977): Geochemistry and origin of phosphorite deposits from off Peru and Chile; *Geol. Soc. Am. Bull.*, **88,** 813-823.

BURNETT W.C. (1980a): Oceanic phosphate deposits; in: *Proc. Fertilizer Raw Material Resources Workshop (1979),* Sheldon R.P. and Burnett W.C. (Eds), 119-144, East-West Centre, Honolulu, Hawaii.

BURNETT W.C. (1980b): Apatite – glauconite associations off Peru and Chile: palaeo-oceanographic implication; *J. Geol. Soc., London,* **137,** 757-764.

BURNETT W.C. and VEEH H.H. (1977): Uranium-series disequilibrium studies in phosphorite nodules from the west coast of South America, *Geochim. Cosmochim. Acta,* **41,** 755-764.

BURNETT W.C. and ROE K.K. (1983): Upwelling and phosphorite formation in the ocean; in: *Coastal Upwelling,* Part A, Suess E. and Thiede J. (Eds), 377-397, Plenum, London.

BURNETT W.C., BEERS M.J. and Roe K.K. (1982): Growth rates of phosphate nodules from the continental margin off Peru; *Science,* **215,** 1616-1618.

BURNETT W.C., VEEH H.H and SOUTAR A. (1980): U-Series, oceanographic and sedimentary evidence in support of recent formations of phosphate nodules off Peru; *SEPM Spec. Publ.,* no. 29, 61-71.

BUROLLET P.F. and OUDIN J.L (1980): Paléocène et Eocène en Tunisie – Pétrole et phosphate; In: *Géologie comparée des gisements de phosphate et de pétrole; Doc. B.R.G.M.,* no. 24, 205-216.

BUSHINSKI G.I. (1964): On shallow water origin of phosphorite sediments; in: *Developments in Sedimentology,* **1,** 62-70.

BUSSON G. (1978): L'unité des faciès confinés en milieu de plate-forme carbonatée; *Doc. Lab. Géol. Fac. Sci. Lyon,* H.S. 4, 87-112.

BUSSON G. and PERTHUISOT J.P. (1977): Intérêt de la Sebkha El Melah (Sud Tunisien) pour l'interprétation des séries évaporitiques anciennes; *Sediment. Geol.,* **19,** 139-164.

CANTO ROMERA J.M. (1978): Aspectos sedimentologicos de la concentracion de fosfatos en el yacimiento de Bu-Craa, Sahara occidental, Tecniterrae, Esp.; *Inst. Geol. Min. Esp., Mapa Geol. Esp.,* **4** (24), 14-30.

CAPDECOMME L. (1952): Etude minéralogique des gîtes de phosphates alumineux de la région de Thiès (Sénégal); *C.R. 19th Congr. Geol. Int., Algiers,* **XI,** 103-107.

CAPDECOMME L. and ORLIAC M. (1967): Sur les caractères chimiques et thermiques des phosphates alumineux de la région de Thiès (Sénégal); *Coll. Int. Phos. Miner. Solides, Toulouse,* 16-20 May 1967, **2,** 45-55.

CARNOT A. (1896): Sur le mode de formation des gîtes sédimentaires de phosphate de chaux; *C.R. Acad. Sci.,* **CXXIII,** 1-6.

CATHCART J.B. (1978): Uranium in phosphate rock; *U.S. Geol. Surv. Prof. Pap.,* **988-A,** 1-6.

CATHCART J.B. and GULBRANDSEN R.A. (1972): Phosphate deposits; *U.S. Geol. Surv. Prof. Pap.,* **820,** 515-525.

CAYEUX L. (1936): Existence de nombreuses bactéries dans les phosphates sédimentaires de tout âge. Consequences; *C.R. Acad. Sci.,* **203,** 1198-1200.

CAYEUX L. (1939,1941,1950): *Les phosphates de chaux sédimentaires de France,* **I, II** and **III,** Etude des gîtes minéraux de la France, *Serv. Carte Géol. Fr.*, (349 pp, 659 pp and 458 pp resp.)

CHAABANI F. (1978): Les phosphorites de la coupe type de Foum Selja (Metlaoui, Tunisie); *Thèse de 3ème cycle,* Université Louis Pasteur de Strasbourg, 27 Oct. 1978.

Chambionnat A. (1963); *Revue de la recherche agronomique marocaine,* **6,** 101-116.

Chamley H. (1971): Recherches sur la sédimentation argileuse en Méditerranée; *Thèse doctorat és sciences,* Université Aix Marseille; *Mém. Sci. Geol.,* **35,** 209 pp.

Chamley H. and Millot G. (1972): Néoformation de montmorillonite à partir de diatomées et de cendres dans les sédiments marins de Santorin (Méditerrannée orientale); *C.R. Acad. Sci., Ser. D,* **274,** 1132-1134.

Champetier Y. (1976): Localisation et approche du processus de formation de l'endogangue des phosphates en débris; *Bull. Soc. Géol. Fr.,* **18** (4), 1017-1021.

Champetier Y. and Joussemet R. (1979a): Enrichissement des phosphates à gangues carbonatées par traitement thermique et lavage. Rapport Action concertée D.G.R.S.T.; *Valorisation des Ressources du sous-sol. Décision d'Aide,* no. 77-7., 1550-1551.

Champetier Y. and Joussemet R. (1979b): Découverte de nubéculaires et d'oncolithes en tant qu'éléments phosphatés; *C.R. Acad. Sci., Ser. D,* **288,** 673-675.

Champetier Y., Blazy P. and Joussemet R. (1980): Enrichissement des phosphates carbonatés. Caractérisation gîtologique et pétrographique. Comportement au traitement thermique; *Proc. 2nd Int. Congr. Phos. Comp. Proc., Boston,* 283-312.

Charles G. (1953): Sur l'origine des gisements de phosphates de chaux sédimentaires; *C.R. 19th Congr. Geol. Int., Algiers,* **XI,** 163-184.

Chateauneuf J.J. (1980): Etude palynologique, stratigraphique et paléoclimatique de l'Eocène supérieur et de l'Oligocène du Bassin de Paris; *Mém. B.R.G.M,* no. 116, 360 pp.

Chien S.H., Wier D.R. and Black C.A. (1975): Supersaturation phenomena and the formation of fluorapatite in aqueous suspensions of phosphate rock; *Soil Sci. Soc. Am., Proc.,* **39,** 43-47.

Christie R.L. (1978): Sedimentary phosphate deposits: an interim review; *Geol. Surv. Can. Pap.,* **78-20,** 1-9.

Christie R.L. (1979): Phosphorite in sedimentary basins of western Canada. Current Research, Part B; *Geol. Surv. Can. Pap.,* **79-IB,** 253-258.

Chu S.P. (1943): The influence of the mineral composition of the medium on the growth of planctonic algae; *J. Ecol.,* **31** (2), 209-148.

Clark R.S. and Altschuler Z.S. (1958): Determination of the oxidation states of uranium in apatite and phosphorite deposits; *Geochim. Cosmochim. Acta,* **13,** 127-142.

Clarke F.W. and Wheeler W.C. (1922): The inorganic constituents of marine invertebrates; *U.S. Geol. Surv. Prof. Pap.,* **124,** 1-62.

CLOCCHIATI R. and SASSI S. (1972): Découverte de témoins d'un volcanisme d'âge paléocène à éocène inférieur dans le bassin phosphaté de Metlaoui (Tunisie méridionale); *C.R. Acad. Sci.,* **274,** 513-515.

COOK P.J. (1967): Winnowing – An important process in the concentration of the Stairway Sandstone (Ordovician) phosphorites of Central Australia; *J. Sediment. Petrol.,* **37** (3), 818-828.

COOK P.J. (1970): Repeated diagenetic calcitization, phosphatization, and silicification in the Phosphoria Formation; *Geol. Soc. Am. Bull.,* **81** (7), 2107-2116.

COOK P.J. (1972): Petrology and Geochemistry of the Phosphate Deposits of Northwest Queensland, Australia; *Econ. Geol.,* **67** (8), 1193 – 1213.

COOK P.J. (1976): Sedimentary Phosphate Deposits; in: *Handbook of Stratabound and Stratiform Ore Deposits,* **7,** 505-535, Elsevier, Amsterdam.

COOK P.J. (1982): World Availability of Phosphorus: An Australian Perspective; in: *Phosphorus in Australia,* Costin A.B. and Williams C.H. (Eds), Monograph 8, 1-41, Centre for Resource and Environmental Studies, Canberra.

COOK P.J. and MCELHINNY M.W. (1979): a Reevaluation of the Spatial and Temporal Distribution of Sedimentary Phosphate Deposits in the light of Plate Tectonics; *Econ. Geol.,* **74,** 315-330.

COOK P.J. and SHERGOLD J.H. (1979): Proterozoic – Cambrian Phosphorites; *1st Int. Field Workshop and Seminar held in Australia, August 1978, Project 156, UNESCO-IUCS,* Canberra Publishing and Printing Co. Pty. Ltd., Canberra, 106 pp.

COPPENS R., BASHIR S. and RICHARD P. (1977): Radioactivity of Al Hasa Phosphate – A preliminary study; *Miner. Deposita,* **12,** 189-196.

COULOMBEAU C. (1979): Une méthodologie des reconnaissance des dépôts phosphatés de la craie sénonienne de Picardie (France); *Chron. Mines Rech. Minière,* no. 499, 5-24.

D'ANGLEJAN B.F. (1967): Origin of marine phosphorites off Baja California, Mexico; *Mar. Geol.,* **5,** 15-44.

DAVEY R.J. (1970): Palynology and palaeo-environmental studies, with special reference to the continental shelf sediments of South Africa; *2nd Plankton. Conf. Proc., Rome, 1971*, 331-347.

DAVIDSON C.F. and ATKIN D. (1953): On the occurrence of uranium of phosphate rock; *C.R. 19th Congr. Geol. Int., Algiers,* **XI,** 13-31.

DAVIES P.J., BUBELA B. and FERGUSON J. (1978): The formation of ooids; *Sedimentology,* **25,** 703-730.

DEFLANDRE G. and TERS M. (1970): Présence de microplancton silurien fixant l'âge des ampélites associés aux phtanites de Brétignolles (Vendée); *C.R. Acad. Sci.,* **270,** 2162-2166.

DICKERTS P. (1973): Use of phosphates in search for oil; *Bull., Am. Assoc. Pet. Geol.,* **57** (2), 432.

DONNOT M. and GUERANGE B. (1978): Le synclinorium cambrien de Brusque. Implications stratigraphiques et structurales dans les monts de l'Est de Lacaune (Tarn, Aveyron, Hérault) – versant nord de la Montagne Noire; *Bull. Bur. Rech. Géol. Min.* Sect. 1, no. 4, 333-363.

DONNOT M., GUIGUES J., LULZAC Y., MAGNIEN A., PARFENOFF A., and PICOT P. (1973): Un nouveau type de gisement d'europium: la monazite grise à europium en nodules dans les schistes paléozoïques de Bretagne; *Mineral. Deposita,* **8,** 7-18.

DUCHATELLE G. (1969): Dosage des espèces minéralogiques renfermant l'ion CO_3= et l'ion SO_4= dans les minerais phosphatés sédimentaires; *Bull. Soc. Chim. Fr.,* Special issue, 1671-1673.

DUMON R. (1971): Evolution et tendance – L'économie mondiale du phosphore et de l'acide phosphorique; *Chim. Ind.–Genie Chim.* **104** (11).

DUNHAM R.J. (1962): Classification of carbonate rocks according to depositional texture. In: *Classification of carbonate rocks – a symposium; Am. Assoc. Pet. Geol. Mem.,* no. 1, 108-121.

DUNOYER DE SEGONZAC G. (1969): Les minéraux argileux dans la diagenèse. Passage au métamorphisme; *Mém. Serv. Carte Géol. Alsace Lorraine,* **29,** 1-320.

EANES E.D. (1975): Amorphous intermediates in the formation of biological apatites; *Coll. Int. C.N.R.S.,* no. 230, 296-301.

EL KAMMAR A.M., ZAYED M.A. and AMER S.A. (1979): Rare earths of the Nile Valley phosphorites, Upper Egypt; *Chem. Geol.,* no. 24, 69-81.

EL MOUNTASSIR M. (1977): La zone rubéfiée de Sidi Dahoui: altération météorique du phosphate de chaux des Ouled Abdoun (Maroc); *Thèse de 3ème cycle,* Université Louis Pasteur, Strasbourg.

ENNEVER J. (1963): Microbiological calcification in comparative biology of calcified tissue; *Ann. N.Y. Acad. Sci.,* **109** (1), 4-13.

ERHART H. (1956): *La genèse des sols en tant que phénomène géologique,* Masson et Cie, Paris.

FAUCONNIER D. (1975): Répartition des péridiniens de l'Albien du Bassin de Paris. Rôle stratigraphique et liaison avec le cadre sédimentologique; *Bull. Bur. Rech. Géol. Min., Sect. 1,* no. 4, 235-273.

FAUCONNIER D. (1979): Les Dinoflagellés de l'Albien et du Cénomanien inférieur du Bassin de Paris. Répartition stratigraphique et relations avec la nature du dépôt; *Thèse,* Université d'Orléans (1977); *Doc. B.R.G.M.,* no. 5, 150 pp.

FAUCONNIER D. and SLANSKY M. (1978): Rôle possible des dinoflagellés dans la sédimentation phosphatée. *Bull. Bur. Rech. Géol. Min. Sect. 4,* no. 3, 191-200.

FAUCONNIER D. and SLANSKY M. (1980): Relations entre le développement des dinoflagellés et la sédimentation phosphatée du bassin de Gafsa (Tunisie). In: *Géologie comparée des gisements de phosphate et de pétrole; Doc. B.R.G.M.*, no. 24, 185-204.

FERRANDO R. (1982): Le phosphore et la vie animale; *C.R. Séances Acad. Agric. Fr.*, **68** (4), 290-307.

FISCHER A.G. and ARTHUR M.A. (1977): Secular variations in the pelagic realm. Deep-water carbonate environments; *SEPM Spec. Publ.*, no. 25, 19-50.

FISCHER A.G. and JEROME D. (1973): Geochemistry of minerals containing phosphorus; in: *Environmental Phosphorus Handbook,* 141-152, Wiley, New York, London, Sydney and Toronto.

FISCHER D.J. and MCCONNEL D. (1969): Aluminium-rich apatite; *Science,* **164,** 551-553.

FLEXER A. and STARINSKY A. (1970): Correlation between phosphate content and the foraminiferal plankton/benthos ratio in chalks (Late Cretaceous, northern Israel); *Sedimentology,* **14** (3-4), 245-258.

FLICOTEAUX R. (1982): Genèse des phosphates alumineux du Sénégal occidental – Etapes et guides de l'altération; *Thèse doctorat ès sciences,* Université Saint-Jérôme, Marseille, 1980; *Mém. Sci. Geol.,* no. 67, 249 pp.

FLICOTEAUX R., NAHON D. and PAQUET H. (1977): Genèse des phosphates alumineux à partir des sédiments argilo-phosphatés du Tertiaire de Lam-Lam (Sénégal); *Sci. Géol. Bull.,* **30** (3), 153-174.

FOLK R.L. (1959): Practical petrographic classification of limestones; *Am. Assoc. Petr. Geol.,* **43** (1), 1-38.

FOURNIE D. (1980): Phosphates et pétroles en Tunisie; in: *Géologie comparée des gisements de phosphate et de pétrole; Doc. B.R.G.M.,* no. 24, 157-166.

FROELICH P.N. (1983): Modern precipitation of carbonate fluorapatite in the upper 20 cm of Peru shelf muds: pore water geochemistry; *Abst. 6th Int. Field-Workshop and Seminar on Phosphorites, Morocco-Senegal*; (Int. Geol. Corr. Prog., Project 156.)

GABSHIN V.M., BOBROV V.A. and ZORKINA L.S. (1974): Quantitative relations between uranium and phosphorus in phosphorites and phosphatic sedimentary rocks; *Lithol. Miner. Resour.,* **9** (6), 740-746.

GEOLOGICAL SURVEY OF IRAN (1968): Recent phosphate discoveries in Iran; Report no. 10, *Geol. Surv. Iran,* 79 pp.

GEORGEL J.M., JACQUEMIN M., RANOUX J., THEBAULT J.Y. and UMBDENSTOCK P. (1962): Essais de recherche de phosphates par scintillomètre aéroporté effectués au Djebel Onk (Algérie); *Bull. Bur. Rech. Géol. Min.,* no. 4, 23-33.

GERVY R. (1973): *Les phosphates et l'agriculture,* Dunod, Paris, 298 pp.

GIMMEL'FARB V.M., KRASIL'NIKOVA A.A. and TUSHINA A.M. (1959): Classification des phosphorites; *Dokl. (Proc.) Acad. Sci.* USSR., **128** (6), 1258-1261.

GONY J.N. (1971): Etude cristallochimique du phosphate uranifère de Bakouma (République centrafricaine). *Commis. Energ. At. [Fr.], Rapp.,* CEA-R-4003.

GULBRANDSEN R.A. (1960): Petrology of the Meade Peak Phosphatic Shale Member of the Phosphoria Formation at Coal Canyon, Wyoming; *U.S. Geol. Surv. Bull.,* **1** (111), 71-146.

GULBRANDSEN R.A. (1966): Chemical composition of phosphorites of the Phosphoria Formation; *Geochim. Cosmochim. Acta,* **30,** 769-778.

GULBRANDSEN R.A. (1969): Physical and chemical factors in the formation of marine apatite; *Econ. Geol.,* **64** (4), 365-382.

GULBRANDSEN R.A. (1970): Relation of carbon dioxide content of apatite of Phosphoria Formation to regional facies; *U.S. Geol. Surv. Prof. Pap.,* **700-B,** B9-B13.

GULBRANDSEN R.A. and CREMER M. (1970): Coprecipitation of carbonate and phosphate from seawater; *U.S. Geol. Surv. Prof. Pap.,* **700 – C,** C125-C126.

GULBRANDSEN R.A. and ROBERSON C.E. (1973): Inorganic phosphorus in seawater; in: *Environmental Phosphorus Handbook,* 117-140, Wiley, New York, London, Sydney and Toronto.

GULBRANDSEN R.A., KRAMER J.R., BEATTY L.B. and MAYS R.E. (1966): Carbonate bearing apatite from Faraday Township, Ontario, Canada; *Am. Mineral.,* **51,** 819-824.

GULBRANDSEN R.A., ROBERSON C.E. and NEIL S.T. (1984): Time and the crystallization of apatite in seawater; *Geochim. Cosmochim. Acta,* **48,** 213-218.

HAMEH P.P. (1983): *Guide de visite du gisement de Taïba;* La Cie Sénégalaise des Phosphates de Taïba, 29 pp.

HEDGES J.I. (1980): Interaction between humic substances and inorganic matter in surface marine sediments. In: *Biogéochimie de la matière organique à l'interface eau-sédiment marin; Coll. Int. C.N.R.S.,* no. 293.

HEIN J.R., SCHOLL D.W., BARRON J.A., JONES M.G. and MILLER J. (1978): Diagenesis of late Cainozoic diatomaceous deposits and formation of the bottom simulating reflector in the southern Bering Sea; *Sedimentology,* **25,** 155-181.

HENIN S. (1982): Le phosphore et la vie – Introduction; *C.R. Séances Acad. Agric. Fr.,* **68** (4), 277-281.

HENRY J.L. (1969): Microorganismes Incertae-Sedis (Acritarches et Chitinozoaires) de l'Ordovicien de la presqu'île de Crozon (Finistère). Gisements de Mort-Anglaise et de Kerglintin; *Bull. Soc. Géol. Minéral. Bretagne,* 59-100.

HERRMANN A.G. (1970): Yttrium and Lanthanides; in: *Handbook of Geochemistry,* Wedepohl K.H. (Ed); vol II – 5, 39, 57-71, Springer-Verlag, Berlin, Heidelberg, New York.

HILLY J. and HAGUENAUER B. (1979): Lorraine Champagne; *Guides géol. régionaux,* Masson et Cie, Paris.

HINZ K., DOSTMANN H. and FRITSCH J. (1982): The Continental Margin of Morocco: Seismic Sequences, Structural Elements and Geological Development; in: *Geology of the Northwest African Continental Margin,* 34-60, Springer-Verlag.

HITE R.J. (1978): Possible genetic relationships between evaporites, phosphorites and iron-rich sediments; *Mt. Geol.,* **14** (3), 97-107.

HURST TH. L. (1958): Manufacture of Elemental Phosphorus and Inorganic Derivatives; in: *Phosphorus and its Compounds,* Van Wazer J. (Ed.), Wiley, New York.

HUTCHINSON G.E. (1950): in: Survey of existing knowledge of biochemistry 3. The biochemistry of vertebrate excretion; *Bull. Am. Mus. Nat. Hist.,* **96**, 554 pp.

ILLYIN A.V. (1975): Connections between dolomite and phosphorite formation in the Khubsugul Phosphorite Basin; *Dokl. Akad. Nauk. S.S.S.R,* **199** (1), 189-191.

ILLYIN A.V. and RATMIKOVA G.I. (1976): Rare earth distribution in the Höbsö Gol. phosphorites (Mongolia); *Geochem. Int.,* **13** (5), 53-56.

INSTITUT MONDIAL DU PHOSPHATE (1980): *Proceedings of the Second International Congress on Phosphate Compounds,* 21-25 April 1980, Boston, U.S.A., IMPHOS, Casablanca, Morocco, 826 pp.

IFA (International Fertilizer Industry Association): *Annual Statistics,* 28 rue Marbeuf, 75008 Paris.

JANSSON M. (1977): Enzymatic release of phosphate in water from subarctic lakes in northern Sweden; *Hydrobiologia,* **56** (2), 175-180.

JEANNETTE D., MONITION A., ORTELLI L. and SALVAN H. (1956): Premiers résultats de l'étude des argiles de la série phosphatée du bassin de Khouribga (Maroc); *20ème Congr. Geol. Int., Mexico, 1956, Comm. Int. Etude Argiles,* 53-60.

KANEL J. (de) and MORSE J.W. (1978): The chemistry of orthophosphate uptake from seawater on to calcite and aragonite; *Geochim. Cosmochim. Acta,* **42,** 1335-1340.

KAZAKOV A.V. (1939): The phosphate facies: origin of the phosphorites and the geologic factors of formation of the deposits; *Proc. Sci. Inst. Fert. Insectofung.,* **145,** 1-106.

KEYSER F. (de) and COOK P.J. (1972): Geology of the Middle Cambrian phosphorites and associated sediments of Northwestern Queensland, *Bur. Miner. Resource. Geol. Geophys. Bull.,* no. 138, 79 pp.

KHARIN G.S. (1974): The phosphatic rocks of the submarine Jan Mayen ridge (North Atlantic) and their relation to vulcanism; *Lithol. Miner. Resour. USA,* **9** (1), 32-35.

KILINC M. and COTILLON P. (1977): Le gisement d'Hahotoé–Kpogamé (Tertiaire du Sud Togo), exemple de piège sédimentaire à sables phosphatés; *Bull. Bur. Rech. Géol. Min., Sect. 2,* no. 1, 43-63.

KOLODNY Y. (1969): Are marine phosphorites forming today?; *Nature,* **224** (5223), 1017-1019.

KRAMER J.R. (1964): Sea water: saturation with apatites and carbonates; *Science,* **146,** 637-638.

KRAUSKOPF B. (1955): Sedimentary deposits of rare metals; *Econ. Geol.,* **50,** 411-463.

KUENZLER E.J. and PERRAS J.P. (1965): Phosphatases of marine algae; *Biol. Bull. (Woods Hole, Mass.),* no. 128, 271-284.

LADOUCE R. (1953): Utilisation des engrais en ostréïculture; *Sci. Peche,* no. 9.

LAFFITTE P. (1977): Engrais et phosphates; *Chron. Mines Rech. Minière.,* no. 435, 3-14.

LAMBOY M. (1976): Géologie marine et sous-marine du plateau continental au Nord-Ouest de l'Espagne. Genèse des glauconies et des phosphorites; *Thèse doctorat ès sciences naturelles,* Université de Rouen.

LAMBOY M. (1982a): Importance des pelotes fécales comme origine des grains de phosphates; l'exemple du gisement de Gafsa (Tunisie); *C.R. Acad. Sci.,* **295**, 595-600.

LAMBOY M. (1982b): La phosphatisation en micro-milieu granulaire d'après un exemple tunisien; *C.R. Acad. Sci.,* **295**, 799-802.

LAMBOY M. and LUCAS J. (1979): Les phosphorites de la marge nord de l'Espagne. Etude géologique et pétrographique; *Oceanol. Acta,* **2** (3), 325-337.

LATRILHE E. (1959): Contribution à l'étude des phosphates alumineux de la région de Thiès (Sénégal); *Bull. Serv. Géol. Prospect. Min. Dakar,* no. 25, 84 pp.

LEHR J.R., MCCLELLAN G.H., SMITH J.P. and FRAZIER A.W. (1967): Characterization of apatites in commercial phosphate rocks; in: *Coll. Int. Phos. Miner. Solides, Toulouse,* 16-20 May 1967, **2,** 29-44.

LOWE D.R. (1971): The relationship between silicic volcanism and the formation of some sedimentary phosphorites; *Proc. VII Forum Geol. Ind. Mineral. Geol. Phosphate, Dolomite, Limestone Clay Deposits, Tampa, Florida,* April 28-30, 217-225.

LOWENSTAM H.A. (1974): Impact of life on chemical and physical process. The Sea; *Mar. Chem.,* **5,** 715-796.

LUCAS J. and NOEL D. (1957): Sur l'importance des coccolithophoridés dans le nannoplancton de phosphates sédimentaires du Lias à l'époque actuelle; *C.R. Acad. Sci.*, **245,** 94-96.

LUCAS J. GROS P. and LANG J. (1977): *Etude microscopique des roches meubles et consolidées;* 503 pp., Doin, Paris.

LUCAS J. and PREVOT L. (1975): Les marges continentales, pièges géochimiques; l'exemple de la marge atlantique de l'Afrique à la limite Crétacé – Tertiaire; *Bull. Soc. Géol. Fr.*, **7** (17-14), 496-501.

LUCAS J. and PREVOT L. (1981): Synthèse d'apatite à partir de matière organique phosphorée (ARN) et de calcite par voie bactérienne; *C.R. Acad. Sci., Ser. B,* **292,** 1203-1208.

LUCAS J. and PREVOT L. (1984): Synthèse de l'apatite par voie bactérienne à partir de matière organique phosphatée et de divers carbonates de calcium dans des eaux douce et marine naturelles; *Chem. Geol.,* **42,** 101-118.

LUCAS J., CHAABANI F. and PREVOT L. (1979a): Phosphorites et évaporites: deux formations de milieux sédimentaires voisins étudiés dans la coupe du Paléogène de Foum Selja (Metlaoui, Tunisie); *Sci. Géol. Bull.,* **32** (7-20).

LUCAS J., PREVOT L., ATAMAN G. and GUNDOGDU N. (1979b): Etude minéralogique et géochimique de la série phosphatée au Sud-Est de la Turquie (Mazidagi-Mardin); *Sci. Géol. Bull.,***32** (1-2), 59-68.

LUCAS J., PREVOT L. and EL. MOUNTASSIR M. (1979c): Les phosphorites rubéfiées de Sidi Daoui. Transformation météorique locale du gisement de phosphate des Ouled Abdoun (Maroc). *Sci. Géol. Bull.,* **32** (1-2), 21-37.

LUCAS J., FLICOTEAUX R., NATHAN Y., PREVOT L. and SHAHAR Y. (1980): Different aspects of phosphorite weathering; *SEPM Spec. Publ.* no. 29, 41-51.

MANSFIELD G.R. (1940): The role of fluorine in phosphate deposition; *Am. Sci.,* **238,** 138-139.

MARTENS C.S. and HARRISS R. (1970): Inhibition of apatite precipitation in the marine environment by magnesium ions; *Geochim. Cosmochim. Acta,* **34,** 621-625.

MARTIN J.H. (1969): Distribution of C, H, N, P, Fe, Mn, Zn, Ca, Sr and Sc in plankton samples collected off Panama and Colombia; *Bioscience,* **19** (10), 898-901.

MASTANDREA O. and LEANZA H.A. (1975): Los Estratos Con Lingula del Norte Argentino y su importancia como fuente de fosfatos; *II Congr. Iberoamericano Geol. Econ., Buenos Aires* **II,** 445-460.

MASTANDREA O., LEANZA H.A., HUGO C.A. and SPIELGELMAN A. (1975): Manifestaciones fosfaticas in la sierra de vaca muerta y adyacencias, Prov. del Neuquen, Rep. Argentina; *II Congr. Iberamericano. Geol. Econ., Buenos-Aires,* **II,** 461-486.

MAUCHLINE J. and TEMPLETON W.L. (1964): Strontium, calcium and barium in marine organisms from the Irish Sea; *J. Cons. Int. Explor. Mer,* **29** (2), 161-170.

MAUGHAN E.K. (1975): Organic Carbon in Shale Beds of the Permian Phosphoria Formation of Eastern Idaho and adjacent states – A summary report; *27th Wyo. Geol. Assoc. Guideb., Ann. Field Conf.,* 107-116.

MAUGHAN E.K. (1980): Relation of phosphorite, organic carbon and hydrocarbons in the Permian Phosphoria Formation, western USA; in: *Geologie comparée des gisements de phosphates et de pétrole; Doc. B.R.G.M.* no. 24, 63-91.

MCARTHUR J.M. (1978): Systematic variations in the contents of Na, Sr, Co_3, and SO_4 in marine carbonate-fluorapatite and their relation to weathering; *Chem. Geol.,* **21**, 89-112.

MCARTHUR J.M. (1980): Post depositional alteration of the carbonate fluorapatite phase of Moroccan phosphates; *SEPM Spec. Publ.,* no. 29, 53-60.

MCCLELLAN G.H. (1980a): Mineralogy of carbonate fluorapatites; *J. Geol. Soc., London,* **137,** 675-682.

MCCLELLAN G.H. (1980b): Quality factors of phosphate raw materials; in: *Fertilizer Mineral Potential in Asia and the Pacific,* Sheldon R.P. and Burnett W.C. (Eds), 359-377, East-West Center, Honolulu, Hawaii.

MCCLELLAN G.H and LEHR J.R. (1969): Crystal chemical investigation of natural apatites; *Am. Mineral.,* **54,** 1374-1391.

MCCONNELL D. (1950): The petrography of rock phosphates; *J. Geol.,* **58,** 16-23.

MCCONNELL D. (1963): Inorganic constituents in the shell of the living brachiopod Lingula; *Geol. Soc. Am. Bull.,* **74,** 363-364.

MCCONNELL D. (1965): Precipitation of phosphates in sea water; *Econ. Geol.,* **60,** 1059-1062.

MCCONNELL D. (1973): *Apatite – its crystal chemistry, mineralogy, utilization and geologic and biologic occurrences,* Springer-Verlag, Vienna and New York, 111 pp.

MCCONNELL D. (1974): The crystal chemistry of apatite; *Bull. Soc. Fr. Mineral. Cristallogr.,* **97,** 237-240.

MCCONNELL D. (1975): Biomineralogy of phosphates and physiological mineralization; in: *Environmental Phosphorus Handbook,* Chap. 25, 425-422, Wiley, New York.

MCKEE E.D., ORIEL S.S. et al. (1967): Paleotectonic Investigations of the Permian System in the United States; *U.S. Geol. Surv. Prof. Pap.,* **515, 217 pp**. (Supplements Paleotectonic Maps of the Permian System.; *Misc. Geol. Invest. Map,* **J-450**.)

McKELVEY V.E. (1963): Successful New Techniques in Prospecting for Phosphate Deposits; *Science, Technology and Development U.S. Pap. for the U.N. Conf. on the application of science and technology for the benefit of less developed areas,* **II**, 163-172.

McKELVEY V.E. (1967): Phosphate deposits. Contribution to economic geology; *Geol. Surv. Am. Bull,* **1252-D**, D1 – D21.

McKELVEY V.E. (1973): Abundance and distribution of phosphorus in the lithosphere; in: *Environmental Phosphorus Handbook,* 13-31, Wiley, New York.

McKELVEY V.E., SWANSON R.W. and SHELDON R.P. (1953): The Permian phosphorite deposits of western United States; *C.R. 19th Congr. Geol. Int., Algiers,* **XI,** 45-64.

McKELVEY V.E., WILLIAMS J.S., SHELDON R.P., CRESSMAN E.R., CHENEY T.M. and SWANSON R.W. (1956): Summary description of Phosphoria, Park City, and Sheldhorn Formations in Western Phosphate Field; *Bull., Am. Assoc. Pet. Geol.,* **40** (12), 2826-2863.

MEHMEL M. (1931): Differences in the crystal structure and chemical formula of apatites; *Z. Phys. Chem.,* **15,** 223-241.

MEYER J.L. and NANCOLLAS G.H. (1973): The influence of multidentate organic phosphorates on the crystal growth of hydroxyapatite, *Calcif. Tissue Res.,* **13,** 205-303.

MIAUTON J.D. (1980): Bakouma – Gélogie et gîtologie d'un gisement néoformé continental phosphato-uranifère; *Thèse doctorat de Spécialité*; Institut National Polytechnique de Lorraine, ENSG, Nancy.

MILLOT G. (1949): Relations entre la constitution et la genèse des roches sédimentaires argile uses; *Thèse* Sci. Nancy et *Géol. Appl. Prospec. Min.,* **2**, no. 2, 3, 4, 1-352.

MILLOT G. (1963): *Geologie des argiles,* Masson et Cie, Paris, 499 pp.

MONCIARDINI CH. (1966): La sédimentation éocène au Sénégal; *Mém. B.R.G.M,* no. 43, 65 pp.

MONTEL G. (1975): Conceptions actuelles sur la structure et la consitution des apatites synthétiques comparables aux apatites biologiques; *Coll. Int. C.N.R.S.,* no. 230, 13-18.

MOORE H.B. (1955): Faecal pellets in relation to marine deposits; in: *Recent marine sediments – a symposium;* Parker D.T. (Ed.), reprinted by Soc. Econ. Paleontol. Mineral., Special publication no. 4, 516-524, Tulsa, Oklahoma, U.S.A.

MOSTAFAVI TABATABAI CH. (1977): Pétrographie et sédimentologie des craies phosphatées du Nord du Bassin de Paris; *Thèse 3ème cycle,* Université de Paris VI.

MUROWCHICK B.L. and WHIPPO R.E. (1967): The crystal chemistry of some sedimentary apatites; *Trans. Soc. Min. Eng.*, no. 238, 257-263.

MURRAY J. and RENARD A.F. (1891): *Deep sea deposits, scientific results of the exploration voyage of H.M.S. Challenger, 1872-1876,* Longman, London., 525 pp.

NARAY-SZABO S.T. (1930): The structure of apatite $(CaF)Ca_4(PO_4)^3$; *Z. Kristallogr., Mineral.*, no. 75, 387-398.

NASH B.D. and BLAKE H.E. (1977): Fluorine recovery from phosphate rock concentrates; *U.S. Bur. Mines, Rep. Invest.*, no. 8205, 1-16.

NATHAN Y. and LUCAS J. (1972): Synthèse de l'apatite à partir du gypse: application au problème de la formation des apatites carbonatées par précipitation directe; *Chem. Geol.*, no. 9, 99-112.

NATHAN Y. and LUCAS J. (1976): Expériences sur la précipitation directe de l'apatite dans l'eau de mer: implication dans la genèse des phosphorites; *Chem. Geol.*, no. 18, 181-186.

NATHAN Y. and SHILONI Y. (1976): Exploration for uranium in phosphorites: a new study on uranium in Israel phosphorites; I.A.E.A., **SM-208/29,** 645-655.

NELSON K.G. and HIGUCHI W.I. (1970): Mechanism of fluoride uptake by hydroxyapatite from acidic fluoride solutions: I. Theoretical considerations; *J. Dent. Res.*, **49** (6), 1541-1548.

NISSENBAUM A. (1979): Phosphorus in marine and non-marine humic substances; *Geochim. Cosmochim. Acta,* **43,** 1973-1978.

NOTHOLT A.J.G. (1978): Proterozoic and Cambrian Phosphorites. Inaugural Field Workshop and Seminar in Australia; *Min. Mag.*, **139 (5)**, 497-499.

NOTHOLT A.J.G. (1979): Resources of Precambrian and Cambrian sedimentary phosphate rock; in: *Proterozoic – Cambrian Phosphorites.* (Cook P.J. and Shergold J.H. (Eds)), 71-86, Canberra Publishing & Printing Co. Pty. Ltd., Canberra.

NOTHOLT A.J.G., Highley D.E. and SLANSKY M. (1979): Raw Materials Research and Development. IV. Dossier on Phosphate; *Comm. Eur. Commun.*, DG XIII, Research, Science, Education, XII/1292/78, 234 pp.

NRIAGU J.O. (1976): Phosphate, clay mineral relation in soil and sediments; *Can. J. Earth Sci.*, **13** (6), 717-736.

NRIAGU J.O. and DELL C.I. (1974): Diagenetic formation of iron phosphates in Recent lake sediments; *Am. Mineral.*, **59,** 934-946.

OBELLIANNE J.M. (1963): Le gisement de phosphate tricalcique de Makatéa (Polynéesie Française); *Sci. Terre,* **IX** (1), 5-60.

ODIN G.S. (1975): Les glauconies; *Thèse de doctorat ès sciences naturelles,* Université Pierre et Marie Curie, Paris.

ODIN G.S. and LETOLLE R. (1980): Glauconitization and phosphatization environments: a tentative comparison; *SEPM Spec. Publ.,* no. 29, 227-237.

PARFITT R.L., ATKINSON R.J. and SMART R. ST. C. (1975): The mechanism of phosphate fixation by iron oxides; *Soil Sci. Soc. Am. Proc.,* **39** (5), 837-841.

PARSONS T.R., STEPHENS K. and STRICKLAND J.D.H. (1961): On the chemical composition of eleven species of marine phytoplankton; *J. Fish. Res. Board Can.,* **18** (4), 1011-1016.

PASCAL M.: The Eocene phosphate deposits of Matam, northeastern Senegal; in: *World Phosphate Rock Resources*, Notholt A.J.G., Sheldon R.P. and Davidson D.F. (Eds), Cambridge University Press, Cambridge. (In press.)

PASHO D.W. (1976): Distribution and morphology of Chatam Rise phosphorites; *N.Z. Oceanogr. Inst. Mem.,* **77**, 27 pp.

PASTERNAK F.A. (1970): Marine biology of the East Atlantic continental margin; in: *Geology of the East Atlantic Continental Margin,* ICSU/SCOR Working Party 31, Symposium, Cambridge, Report no. 70/13, *Inst. Geol. Sci.,* 66-67.

PATTON W.W., Jr and MATZKO J.J. (1959): Phosphate deposits in northern Alaska; *U.S. Geol. Surv. Prof. Pap.,* **302-A**, 17 pp.

PATWARDHAN A.M. (1978): Permian fossils and stromatolitic structures from Mussoorie phosphorites, Lower Himalaya, India, *Abstracts Int. Sedim. Congress, Jerusalem.*

PATWARDHAN A.M. and AHLUWALIA A.D. (1979): A note on the original Mussoorie phosphorite in the Lower Himalaya, India, and its paleogeographic implications; *Miner. Deposita,* **8** (4), 379-387.

PEVEAR D.R. (1966): The estuarine formation of United States Atlantic coastal plain phosphorite; *Econ. Geol.,* **61** (2), 251-256.

PENEAUD P. (1978): La paragenèse phosphatée de la grotte de la caune de l'Arago (Pyrénées orientales); *Thèse de doctorat de 3ème cycle,* Université de Paris VII.

PERES J.M. (1963): *Océanographie biologique et biologie marine. Tome 1: la vie benthique,* Presses Univ. de France, Paris, 514 pp.

PERES J.M. (1976): *Précis d'océanographie biologique,* Presses Univ. de France, Paris. 246 pp.

PERES J.M. and DEVEZE L. (1964): *Océanographie biologie et biologie marine. Tome 2. La vie pélagique,* Presses Univ. de France, Paris.

PETTIJOHN F.J. (1957): *Sedimentary rocks,* 2nd edn, Harper and Row, New York, Evanston and London, 718 pp.

PIPER D.Z. and CODISPOTI L.A. (1975): Marine phosphorite deposits and the nitrogen cycle; *Science,* **188** (4183), 15-18.

PIRLET H. (1965): Classification des calcaires de la série des arénocryptites; *Ann. Soc. Géol. Belg.,* **88** (7/10), 395-410.

PORTER K.G. and ROBBINS E.I. (1981): Zooplankton fecal pellets link fossil fuel and phosphate deposits; *Science,* **212** (4497), 931-933.

POUTANEN E.L. and MORRIS R.J. (1983): The occurrence of high molecular weight humic compounds in the organic-rich sediments of the Peru continental shelf; *Oceanol. Acta,* **6** (1), 21-28.

POWEL T.G., COOK P.J. and MCKIRDY D.N. (1975): Organic geochemistry of phosphorites: relevance to petroleum genesis; *Bull., Am. Assoc. Pet. Geol.,* **59** (4), 618-632.

PREVOT L. and LUCAS J. (1979): Comportement de quelques éléments traces dans les phosphorites; *Sci. Géol. Bull.,* **32** (1-2), 91-105.

PREVOT L. and LUCAS J. (1980): Behavior of some trace elements in phosphatic sedimentary formations; *SEPM Spec. Publ.,* no. 29, 31-39.

PRIAN J.P. (1980): Caractérisation des paléo-environments des phosphorites cambriennes du versant septentrional de la Montagne Noire (Sud du Massif Central, France); in: *Géologie comparée des gisements de phosphate et de pétrole; Doc. B.R.G.M.,* no. 24, 93-111.

PRIAN J.P.: The Farim-Saliquinhe Eocene phosphate deposit, Guinea – Bissau, West Africa; in: *World Phosphate Rock Resources,* Notholt A.J.G., Sheldon R.P. and Davidson D.F. (Eds), Cambridge University Press, Cambridge (In press.)

PRICE N.B. and CALVERT S.E. (1978): The geochemistry of phosphorites from the Namibian Shelf; *Chem. Geol.,* **23,** 151-170.

RACHIDI M. (1983): Contribution à la connaissance de la matière organique associée à certaines séries phosphatées, *Thèse de doctorat de spécialité,* Université d'Orléans, France.

RADIER H. (1957): Contribution à l'étude géologique du Soudan Oriental (A.O.F.) *Bull. Serv. Géol. Prospect. Min. Dakar,* no. 26, vol 2, 315-556.

RANCHIN G. (1963): Gisement de phosphate de chaux sédimentaire du Djebel Onk (Algérie); *S.E.R.M.I.,* Paris, 85 pp.

RANKE U., VON RAD U. and WISSMANN G. (1982): Stratigraphy, facies and development of the on- and offshore Aaiun-Tarfaya Basin. A Review; in: *Geology of the Northwest African Continental Margin,* 86-105, Springer-Verlag, Berlin, Heidelburg and New York.

REEVES N.J. and SAADI T.A.K. (1971): Factors controlling the deposition of some phosphate bearing strata from Jordan; *Econ. Geol.,* **66,** 451-465.

RIGGS S.R. (1979a): Petrology of the Tertiary Phosphorite System of Florida; *Econ. Geol.,* **74,** 195-220.

RIGGS S.R. (1979b): Phosphorite sedimentation in Florida – A Model Phosphogenic System; *J. Geol. Soc. London,* **137**, 285-314.

RILEY G.A., STOMMEL M. and BUMPUS D.F. (1949): Qualitative ecology of the plankton of the western North-Atlantic; *Yale Univ. Oceanogr. Lab.*, no. 12, 1-69.

ROBERSON C.E. (1966): Solubility implications of apatite in sea water; *U.S. Geol. Surv. Prof. Pap.*, **550-D,** D178-D185.

ROGERS J.K. and KEEVERS R.E. (1976): Lady Annie–Lady Jane Phosphate deposits, Georgina basin, Queensland; in: *Economic Geology of Australia and Papua New Guinea. 4. Industrial Minerals and Rocks,* Knight C.L (Ed), *Monogr. Ser.* no 8, *Australas. Inst. Min. Metall.*, **4**, 251-265.

ROMANKEVICH Y.A. and BATURIN G.N. (1972): Composition of the organic matter in phosphorites from the continental shelf of southwest Africa. *Geochem. Int.*, **9** (3), 464-470.

RONOV A.B. and YAROSHEVSKY A.A. (1969): Chemical Composition of the Earth's Crust; in: *The Earth's Crust and Upper Mantle,* Hart P.J. (Ed.), **13,** 37-57, American Geophysical Union.

ROSENQUIST I.TH. (1970): Formation of vivianite in Holocene clay sediments; *Lithos,* **3** (4).

RUSSEL R.T. and TRUEMAN N.A. (1971): The geology of the Duchess phosphate deposits, Northwestern Queensland, Australia; *Econ. Geol.*, **66** (8), 1186-1214.

SAINT-GUILHEM R. (1960): Les gisements de phosphates de chaux dans le monde; *Notes Maroc.,* no. 14, 21-38.

SALAS G.P. (1978): Sedimentary phosphate deposits in Baja California, Mexico; *AIME Ann. Meet. Denver, Colorado.* Preprint no. 78-H-75, 21 pp.

SALEHI SIAVOCHANI H. (1978): Etude statistique, pétrographie, sédimentologique et géochimique du Paléocène et de l'Eocène phosphaté au N.NE et au centre de la Tunisie; *Thèse de 3ème cycle,* Université de Paris VI.

SALVAN H. (1944): A propos des formations siliceuses des phosphates marocains; *Geol. Rundsch.,* **43** (2), 503-515.

SALVAN H. (1954): Les invertébrés fossiles des phosphates marocains; *Notes Mém. Dir. Mines Géol. Maroc,* no. 93, 257 pp.

SALVAN H. (1959): Le problème de la phosphatogénèse et son évolution; *Mines Géol. Rabat*, no. 6, 31-41.

SAMB M. (1983): Typologie et caractérisation physico-chimique des particules phosphatées – application à la série phosphatée eocène de Casamance (Sénégal); *Thèse docteur ingénieur,* Université d'Orléans, 227 pp.

SANDSTROM M.W. (1982): Organic geochemistry of phosphorites and associated sediments, *PhD Thesis,* University of Canberra.

SARASWAT A.C., SANKARAN A.V., VARADA RAJU M.N., TANEJA P.C. and BARGAJE V.G. 1970: *Geochemical data on the uraniferous phosphorites of Mussoorie, Dehradun district, Uttar Pradesh, India*; in: Proceedings of the Second Seminar on Geochemical Prospecting Methods and techniques; *Miner. Resour. Dev. Ser.,* no. 38, 179-187, United Nations, New York.

SASSI S. (1974): La sédimentation phosphatée au Paléocène dans le Sud et le Centre-Ouest de la Tunisie; *Thèse doctorat ès sciences,* Université de Paris-Sud.

SASSI S. (1980): Contexte paléogéographique des dépôts phosphatés en Tunisie;in: *Géologie comparée des gisements de phosphate et de pétrole*; *Doc. B.R.G.M.,* no. 24, 167-183.

SAVENKO V.S. (1978): Experimental studies on the conditions of chemical precipitation of calcium phosphates from seawater; *Dokl. Akad. Nauk. S.S.S.R,* **243** (5), 1302-1305.

SCOLARI G. and LILLE R. (1973): Nomenclature et classification des roches sédimentaires (Roches détritiques terrigènes et roches carbonatées); *Bull. Bur. Rech. Géol. Min., Sect. 4,* no. 2, 57-132.

SEMENOV E.J., KHOLODOV V.N. and BARINSKIJ R.L. (1962): Rare earths in phosphorites, *Geochim. int.,* **5**, 501-507.

SHAPIRO L. (1952): Simple field method for the determination of phosphate in phosphate rocks; *Am. Mineral.,* **37** (3-4), 341-342.

SHELDON R.P. (1959): Geochemistry of uranium in phosphorites and black shales of the Phosphoria Formation; *U.S. Geol. Surv. Bull.,* **1984-D**, 83-115.

SHELDON R.P. (1964): Paleolatitudinal and paleogeographic distribution of phosphorites; *U.S. Geol. Surv. Prof. Pap.,* **501 C,** 106-113.

SHELDON R.P. (1966): Long distance migration of oil in Wyoming; *Mt. Geol.,* **4** (2), 53-65.

SHELDON R.P. (1980): Episodicity of phosphate deposition and deep ocean circulation. A hypothesis; *SEPM Spec. Publ.,* no. 29, 239-247.

SHELDON R.P. (1981): Ancient marine phosphorites; *Ann. Rev. Earth Planet. Sci.,* **9,** 251-284.

SHELDON R.P. and BURNETT W.C. (Eds) (1980): Fertilizer Mineral Potential in Asia and the Pacific; *Proc. Fert. Raw Mater. Res. Workshop, Aug. 20-24, 1979,* 481 pp., East-West Center, Honolulu, Hawaii.

SHELDON R.P., MAUGHAN E.K. and CRESSMAN E.R. (1968): Sedimentation of rocks of Leonard (Permian) age in Wyoming and adjacent states; in: *Anatomy of the Western Phosphate Field, Intermt. Assoc. Geol., 15th Ann. Field Conf*, 1-13.

SHISHKINA O.V. (1966): Fluorine in oceanic sediments and their interstitial waters; *Geokhimiya*, no. 2, 236-243.

SHOLKOVITZ E. (1973): Interstitial water chemistry of the Santa Barbara Basin sediments; *Geochim. Cosmochim. Acta,* **37,** 2043-2073.

SICKO-GOALD L.M., GRANG R.E. and JENSEN T.E. (1975): Phosphate metabolism in blue-green algae. IV. In situ analysis of polyphosphate bodies by X-ray energy dispersive analysis; *Cytobiologie Z. Exp. Zellforsch,* **11** (3), 430-437.

SLANSKY M. (1962): Contribution à l'étude géologique du bassin sédimentaire côtier du Dahomey et du Togo; *Mém. B.R.G.M,* no. 11, 270 pp.

SLANSKY M. (1977): *Disponibilités et besoins futurs en minerais phosphatés compte tenu de la lutte anti-pollution et de nouvelles applications possibles,* Study carried out for the European Economic Commission in 1975, B.R.G.M., Orléans, 68 pp.

SLANSKY M. (1979a): Proposals for nomenclature and classification of sedimentary phosphate rocks.; in: *Proterozoic-Cambrian Phosphorites,* Cook P.J. and Shergold J.H. (Eds), 60-63, Canberra Publishing & Printing Co. Pty. Ltd., Canberra.

SLANSKY M. (1979b): Répartition et possibilités de concentration de l'uranium dans les phosphates sédimentaires; in: *Uranium deposits in Africa: geology and exploration,* IAEA-AF-**109/14,** 239-249.

SLANSKY M. (1980a): Géologie des phosphates sédimentaires; *Mém. B.R.G.M.,* no. 114, 92 pp.

SLANSKY M. (1980b): Ancient upwelling models: Upper Cretaceous and Eocene phosphorite deposits around West Africa; in: *Fertilizer mineral potential in Asia and the Pacific,* Sheldon R.P. and Burnett W.C. (Eds), 145-148, East-West Center, Honolulu, Hawaii.

SLANSKY M., CAMEZ TH. and MILLOT G. (1959): Sédimentation argileuse et phosphatée au Dahomey; *Bull. Soc. Géol. Fr.,* I, 150-155.

SLANSKY M. and FAUCONNIER D. (1973): Influence possible de certains facteurs biologiques sur la géochimie des sédiments; *Bull. Bur. Rech. Géol. Min. Sect. 4,* no. 4, 209-228.

SLANSKY M., LALLEMAND A. and MILLOT G. (1965): La sédimentation et l'altération latéritique des formations phosphatées du gisement de Taïba (Sénégal); *Bull. Serv. Carte Géol. Alsace Lorraine,* **17** (4), 311-324.

SMIRNOV A.I., IVNITZKAYA R.B. and ZALAVINA T.P. (1962): Experimental data on the possibility of the chemical precipitation of phosphate from sea water; in: *Geology of phosphorite deposits,* Yostoptekhizdat, Moscow (translated in *State Sci. Res. Inst. Chem. Raw Mater.,* **7,** 289-302).

SOUDRY D. (1979): Intervention de schizophytes dans la phosphimicritisation des débris osseux; *C.R. Acad. Sci. Ser. D,* **288,** 669-671.

SOUDRY D. and CHAMPETIER Y. (1983): Microbial processes in the Negev phosphorites (southern Israel); *Sedimentology,* **30,** 411-423.

SOUDRY D. and NATHAN Y. (1980): Phosphate peloids from the Negev phosphorites; *J. Geol. Soc., London,* **137,** 749-755.

SOVIET GEOLOGICAL DICTIONARY (1973): *Geologicheskiji slovar',* two volumes, Nedra, Moskow.

STETS J. and WURSTER P. (1982): Atlas and Atlantic – Structural Relations; in: *Geology of the Northwest African Continental Margin,* 69-85, Springer Verlag, Berlin, Heidelberg and New York.

STOW S.H. (1976): Effects of weathering on the chemical and heavy mineral composition and physical properties of phosphate pebbles from the Bone Valley Formation of Florida; *Southeast. Geol.,* **18** (2), 83-98.

STOWASSER W.F. (1983): *Phosphate Rock Mineral Commodity Profiles, US Bur. Mines,* 18 pp.

STRAKHOV N.M. (1969): *Principles of Lithogenesis,* **2**, 235-261, Oliver and Boyd, Edinburgh.

STUMM W. and LECKIE J.O. (1970): Phosphate exchange with sediments: its role in the productivity of surface waters; in: *Advances in Water Pollution Research. 2,* **III,** 26/16, Pergamon Press, Oxford.

SUESS E. and FUTTERER R. (1972): Aragonitic ooids: experimental precipitation from seawater in the presence of humic acid; *Sedimentology,* **19,** 129-139.

SUMMERHAYES C.P. (1973): Distribution, origin and economic potential of phosphatic sediments from the Agulhas bank, South Africa; *Trans. Geol. Soc. S. Afr.,* **76** (3), 271-277.

SWANSON R.W. (1970): Mineral resources in Permian Rocks of Southwest Montana *U.S. Geol. Surv. Prof. Pap.,* **313-E**, 661-777.

SYERS J.K., SHAH R. and WALKER T.W. (1969): Fractionation of phosphorus in two alluvial soils and particle-size separates; *Soil Sci.,* **108** (4), 283-289.

SYERS J.K., WILLIAMS J.D.H. and WALKER T.W. (1970): Mineralogy and forms of inorganic phosphorus in a greywacke soil-rock weathering sequence; *Soil Sci.,* **110** (2), 100-106.

TARDIEUX-ROCHE A. (1966): Contribution à l'étude des interactions entre phosphates naturels et microflore du sol; *Thèse sciences naturelles,* 132 pp., Inst. Nat. Rech. Agron., Paris.

TESSIER F. (1952): Contribution à la stratigraphie et à la paléontologie de la partie ouest du Sénégal (Crétacé et Tertiaire); *Bull. Dir. Mines Géol. Dakar,* no. 14, vols I and II, 571 pp.

TEODOROVICH G.I. (1947): Sedimentary geochemical facies; *Byull. Mosk. Ova. Ispyt. Prir., Otd. Geol.,* **22** (1).

TISSOT B. and WELTE D.H. (1978): *Petroleum Formation and Occurrence, a New Approach to Oil and Gas Exploration,* 538 pp. Springer-Verlag, – Berlin.

TOOMS J.S., SUMMERHAYES C.R. and CRONAN D.S. (1969): Geochemistry of marine phosphate and manganese deposits; *Oceanogr Mar. Biol. Ann. Rev.,* **71**, 49-100.

UNITED NATIONS (1968): Proceedings of the seminar on source of mineral raw materials for the fertilizer industry in Asia and the Far East, Bangkok 1967; *Miner. Resour. Dev. Ser.,* no. 32, 392 pp., New York.

UNITED NATIONS (1980): Fertilizer Manual; *Dev. Transf. Technol. Ser.,* no. 13, 353 pp., New York.

U.S. BUREAU OF MINES – U.S GEOLOGICAL SURVEY (1981): Principles of a resource/reserve classification for minerals; *U.S. Geol. Surv. Circ.,* **831,** 5 pp.

VAN BENNEKOM A.J. and VAN DER GAAST S.J. (1976): Possible clay structures in frustules of living diatoms; *Geochim. Cosmochim. Acta,* **40,** 1149-1152.

VATAN A. (1959): La sédimentation et les sédiments marins actuels; *Mines Géol. Rabat,* no. 6.

VEEH H.H., BURNETT W.C. and SOUTAR A. (1973): Contemporary phosphorites on the continental margin of Peru; *Science,* **181**, 845-847.

VEEH H.H., CALVERT S.E. and PRICE N.B. (1974): Accumulation of uranium in sediments and phosphorites on the South West African shelf; *Mar. Chem.,* **2,** 189-202.

VINOGRADOV A.P. (1953): *The elementary chemical composition of marine organisms,* translated by J. Efron and J.K. Setlow, Sears Foundation for Marine Research, Yale University, New Haven.

VISSE L.D. (1953): Genèse des gîtes phosphatés du sud-est algéro-tunisien; *19th Int. Geol. Congr. Algiers,* ser. 1, **27**, 53 pp.

VISSE L.D. (1953): Les faciès phosphatés; *Rev. Inst. Fr. Pét.,* Special Issue, 87-98.

VISSE L.D. (1974): Digénie des phosphates en grains et ses conséquences; *C.R.Acad. Sci.,* **278** (24), 3051-3054.

VON RAD U. and WISSMANN G. (1982): Cretaceous – Cenozoic History of the West Saharan Continental Margin (NW Africa): Development, Destruction and Gravitational Sedimentation; in: *Geology of the Northwest African Continental Margin,* 86-105, Springer-Verlag, Berlin, Heidelberg and New York.

VON RAD U., HINZ K., SARNTHEIN M. and SEIBOLD E. (Eds) (1982): *Geology of the Northwest African Continental Margin,* 703 pp., Springer-Verlag, Berlin, Heidelberg and New York.

WALKER T.W. (1972): Soil phosphorus and its fate; Proc. Press. Forum Eutroph., 33-42.

WELLS F.J. (1975): *The Long-run Availability of Phosphorus,* 121 pp., J. Hopkins University Press, Baltimore and London.

WESTIN FRED C. and DE BRITO JULIA G. (1969): Phosphorus fractions of some Venezuelan soils as related to their stage of weathering; *Soil Sci.,* **107** (3), 194-202.

WILLIAMS D.B. (1971): *The occurrence of dinoflagellates in marine sediments: the micropaleontology of oceans,* Funnel and Riedel (Eds), 231-243, Cambridge University Press, Cambridge.

WILLIAMS J.D.H. and WALKER T.W. (1969): Fractionation of phosphate in a maturity sequence of New Zealand basaltic soil profiles: 2; *Soil Sci.,* **107** (3), 213-219.

WILLIAMS J.D.H., SYERS J.K., WALKER T.W. and SHAH R. (1969): Apatite transformations during soil development; *Agrochimica,* **XIII** (6), 491-501.

WINNACKER K. and KUCHLER L. (1964): *Traité de chimie appliquée – Technologie minérale,* Part 2, 452 pp., Editions Eyrolles, Paris.

WOBBER F.J. and MARTIN K. (1977): Remote sensing applications to P_2O_5 mining and environmental planning; *World Min.,* **30** (4), 53-57.

WOLLAST R. and DE BROEU F. (1971): Study of the behaviour of dissolved silica in the estuary of the Scheldt; *Geochim. Cosmochim. Acta,* **35,** 613-620.

WORTHINGTON L.V. (1958): Oceanographic data from the R.R.S. Discovery II, International Geophysical Cruises 1 and 2, 1957; Woods Hole Oceanographic Inst., WHOI no 58-30, 84 pp.

YOUNG R.A. (1967): Dependence of apatite properties on crystal structural details; *Trans. N.Y. Acad. Sci.,* **29** (II), 949-959.

YOUNG R.A. (1975): Some aspects of crystal structural modeling of biological apatites; in: *Physico-chime et cristallographie des apatites d'intérêt biologique,* 21-40, C.N.R.S., Paris.

YOUSSEF M.I. (1965): Genesis of bedded phosphates; *Econ. Geol.,* **60,** 590-600.

ZAGHLOUL Z.M. and MABROUK B. (1964): On uranium in Dakhla and Mahamid phosphate deposit; *J. Geol. U.A.R.,* **8** (2), 79-87.

INDEX

Accumulation trap, 134-36, 149
ADP, 14
Algae, 89, 93, 111
Aluminium, 73, 141
Aluminium phosphate minerals, 33
Aluminium phosphorites, 139-40, 161
Aluminocalcic phosphates, 35, 141
Aluminophosphatites, 46, 169
 definition, 39
Ammonites, 66, 115
Ammonium phosphates, 34
AMP, 14
Apatite lattice, 27, 29, 30
Apatites, 19-33, 52
 cryptocrystalline, 41
 crystalline textures, 41
 evaluation of substitutions, 31-32
 formation of sedimentary, 98-103
 influence of structure on substitutions, 26-28
 influence of substitution of PO_4 by CO_3 on physical properties, 32-33
 influence of substitutions on structure, 28-31
 lattice parameters, 30, 31
 macrocrystalline, 41
 megacrystalline, 41
 other frequent substitutions, 22
 principal substitutions, 19-22
 refractive index, 33
 relations between structure and substitutions, 23-32
 sedimentary, factors triggering formation of, 103-12
 structure of, 22-23
ATP, 14
Attapulgite, 67, 102, 105
Augelite, 34
Autunite, 35

Bacteria, 111, 112
Barrandite, 35
Baryte, 72
Benin-Togo coastal basin, 144-48
Benthic organisms, 130
Bioclasts, 40
Biogenesis, 98, 103
Biopelphospharenite, 43, 54
Biophospharenite, 44
Biophospharudite, 43
Black shale, 130
Bofal deposit, 149-51
Bou Craa deposit, 154
Boyle, Robert, 13
Brand, Henning, 13
Brown clay, 139-40

Cacoxenite, 34
Cadmium, 179
Calcite, 67, 101, 107
Calcium, 102
Calcium phosphates, 13, 19, 33, 34, 89, 100, 105, 113, 141, 169
Carbonate fluorapatites, 31-32, 102, 110
Carbonic anhydrase, 111
Cattle feed, 177-78
Celesite, 72
Chalcedony, 33
Chemistry of phosphatites, 69-81
Cherts, 124, 125, 148
Chlorine, 72, 167
Clay, 148
Clay minerals, 101, 102, 124
Clinoptilolite, 105
Collophanite, 19
Conodonts, 98
Coproliths, 41, 98, 117, 120
Crandallite, 34, 35, 46
Crystallographic parameters, 31-32

Denitrification processes, 130
Diatomaceous muds, 94, 95
Diatoms, 92, 102, 124
Dinoflagellates, 92, 102, 124
Direct precipitation, 99-103, 118, 120
DNA, 14

Dolomite, 32, 52, 67, 101, 102

Economic data, 169-71
Enzymes, 111
Epigenesis, 103, 113, 115
Exogangues, 61-67, 137
Extraction methods, 166-67

Faecal pellets, 41, 117
Farim deposit, 149, 152
Favourable zones,
  detailed study of, 165-66
  identification of, 165
Feldspar, 52, 63
Fertilizers, 179
  for direct application, 176
  manufacture of, 174-75
  sintered phosphates, 176
Fluorapatite, 22, 33, 137
Fluorine, 34, 72, 102, 137
Foraminifera, 50, 56, 58
Foucherite, 34
Francolite, 103
Fucus vesiculosus, 89
Fused phosophates, 176

Gabon basin, 142-44
Gafsa basin, 157
Gambia, 148-53
Gangue, 137
Ganntour deposit, 154
Gastropods, 66, 115
Geochemical prospecting, 165
Geological periods, 163-65
Glauconite, 50, 67
Gorceixite, 34
Goyazite, 34
Guano, 33, 34
Guinea, 148-53
Gymnodinium brevis, 92
Gypsum, 105

Humic acids, 96, 98, 110-11, 119, 123
Humic compounds, 96
Hyalinacia artifex, 16
Hydroxyapatite, 112

Igneous deposits, 18
Illite, 67
Imports, 169
Intraclasts, 41
Iron, 73

Kaolinite, 67
Kasserine Island, 154-57
Kerogen, 96
Kunckel, Johann, 13

Lanthanides in phosphatites, 80-81
La'Youn basin, 153
Leaching, 137, 140
Limestones, 52, 105
Lipscombite, 34
Lithoclasts, 41
Lithophospharudite, 43, 44

Magnesium, 88, 101, 102, 112, 119
Magnesium carbonate, 105
Magnesium phosphate, 13, 34
Marcasite, 72
Marine organic matter, 97
Marine solutions, phosphate minerals from,
  88-120
Mauritania-Senegal-Gambia-Guinea Bissau
  basin,
  148-53
Meskala deposit, 154
Meta-autunite, 35
Microsphatites, 46-47
Millisite, 34, 46
Mineralogy of sedimentary phosphates, 19-35
Mining, 166-68
Mitridatite, 34
Monazite, 35
Monetite, 34
Montgomeryite, 35
Moroccan coastal basins, 153-54

N'Diendouri deposit, 149-51
Nitrogen, 130
Nucleic acids, 13-14

Onuphis tubicola, 16
Ooliths, 40, 42, 56, 65, 119
Oophospharenite, 42
Oosphatite, 42
Opal, 101, 102, 107
Ore deposits, 120-36
Ore dressing, 167
Organic carbon, 101, 102
Oulad-Abdoun deposit, 154

Pelecypods, 66, 115
Pellets, 40
Pelphosphalutite, 43
Pelphospharenite, 43, 56
Petrographic classification, 37-46
Petrography of sedimentary phosphates, 37-67
Petroleum, 107
Phosphagene, 14
Phosphalutite, 42
Phospharenite grains, 50, 52
Phospharenites, 42, 43, 44, 47, 61, 63, 117
Phospharenolutite, 42
Phospharenorudite, 42

Phosphatases, 14
Phosphate deposits, 179
  favourable geological periods, 125-31, 163-65
  formation of, 18
  identification of favourable zones, 165
  location of, 134-36
  secondary evolution of, 136-40
  weathering of, 137-38, 161
Phosphate formation, summary of mechanisms of, 159-61
Phosphate industry, 169-78
Phosphate mineralization, 121
Phosphate minerals,
  environment for formation of, 93-98
  formation from marine solutions, 88-120
  non-apatitic, 33-35
Phosphate nodules, 41, 65, 95, 96, 136
Phosphate particles, 42, 47-61, 112-20, 137
  definition, 39-41
  of arenite size, 50-56
  of rudite size, 56-61
Phosphate pebbles, 67
Phosphate pellets, 52, 113, 117
Phosphate reserves, 173-74
Phosphate resources, 171-73
Phosphate rocks, 179-80
  definition, 37-39
Phosphate sedimentation, 83-161
  characteristics of, 159-61
  degree of, 127
  rate of, 120-23
Phosphate supply, abundance of, 131-33
Phosphates,
  consumption of, 13
  definition, 37
  for agriculture, 13
  natural, 13, 19
  uses of, 174-78
Phosphatite accumulation,
  mechanisms of, 83-161
  regions favourable to, 131-36
Phosphatite deposits, 120
Phosphatite nomenclature, 42-44
Phosphatite sedimentation, 124-25
Phosphatites, 63
  and uranium, 79-80
  chemistry of 69-81
  classification of 44-45
  definition, 38-39
  environments favourable to accumulation of, 142-61
  lanthanides in, 80-81
  major elements, 69-73
  phosphate fraction forms, 46-60
  rare earths in, 80-81
  scandium in, 80-81
  trace elements, 73-75
  uranium in, 75-80
  yttrium in, 80-81
Phospho-amino lipids, 14
Phosphoglyceric acids, 14
Phosphorites,
  definition, 37-38, 39
  uranium in, 140-42
Phosphorus,
  concentration by marine organisms, 89-92
  direct continental supply, 83-86
  distribution in nature, 13-18
  earliest extractions, 13
  earth's crust, 17
  essential organic forms found in nature, 1[illegible]
  historical note, 13
  in bone tissue, 13
  in human body, 14
  in living cells, 13
  in marine animals, 16
  in plants, 16
  in sedimentary rocks, 17
  in soil, 85
  in soil and seas, 17
  possible sources of, 83-88
  supply from oceanic waters, 86-88
Phytoplankton, 89, 92, 102
Pisolites, 40
Plankton, 105, 124, 133
Polyphosphates, 178
Price variations, 171
Processing methods, 167-68
Prospecting methods, 163-66
Pseudo-ooliths, 40
Pycnocline, 105
Pyrites, 72

Quartz, 52, 63

Rare earths in phosphatites, 80-81
Red tide phenomenon, 104
RNA, 14
Rockbridgite, 34
Rocky Mountains, 157-59

Sardinia caerula, 105
Scandium in phosphatites, 80-81
Scheele, Carl Wilhem, 13
Scintillometers, 164
Seawater,
  phosphorous in, 86-88
  supersaturation of, 89
Sedimentary deposits, 18
Senegal, 148-53
Sepiolite, 67, 102, 105
Silica, 105, 124
Silicates, 105, 124
Siliceous rocks, 35
Silicon, 102

Sintered phosphates, 176
Smectites, 67, 105
Soils, phosphorus in, 17, 85
Souar formation, 156
Sponges, 124
Strengite, 34, 35
Sulphates, 72
Sulphides, 72
Sulphur, 72

Tarfaya basin, 153
Temora longicornis, 16
Thermal phosphoric acid, 177
Thermal phosphorus, 177
Thermophos process, 176
Thysanoessa inermis, 16
Tinticite, 34
Tunisia, 154-57
Turquoise, 34

Upwelling currents, 131-32, 159
Uranium, 35
 in phosphatites, 75-80
 in phosphorites, 140-42

Variscite, 35
Vivianite, 34
Volcanism, 104

Wavellite, 34
Weathering, 142
 of phosphate deposits, 137-38, 161
Whitlockite, 34
World consumption, 179
World output, 169
World production, 170
World resources and reserves, 171-73

X-ray diffractograms, 31

Yttrium in phosphatites, 80-81